W0235378

CHOOSING & USING
the Right
METAL SHOP LATHE

CHOOSING & USING
the Right
METAL SHOP LATHE

RICHARD REX

INDUSTRIAL PRESS, INC.

Industrial Press, Inc.

32 Haviland Street, Suite 3
South Norwalk, Connecticut 06854
Phone: 203-956-5593
Toll-Free in USA: 888-528-7852
Email: info@industrialpress.com

Author: Richard Rex
Title: Choosing & Using the Right Metal Shop Lathe
Library of Congress Control Number: 2022936409

© by Industrial Press, Inc.
All rights reserved. Published in 2022.
Printed in the United States of America.

ISBN (print):   978-0-8311-3681-9
ISBN (ePUB):   978-0-8311-9616-5
ISBN (eMOBI): 978-0-8311-9617-2
ISBN (ePDF):   978-0-8311-9615-8

Publisher/Editorial Director: Judy Bass
Copy Editor: Judy Duguid
Compositor: Patricia Wallenburg, TypeWriting
Proofreader: David Johnstone
Indexer: Claire Splan

No part of this book may be reproduced or transmitted in any form or by any means, electronic or mechanical, including photocopying, recording, or by any information storage and retrieval system, without written permission from the publisher.

Limits of Liability and Disclaimer of Warranty
The author and publisher make no warranty of any kind, expressed or implied, with regard to the documentation contained in this book.
All rights reserved.

books.industrialpress.com
ebooks.industrialpress.com
1 2 3 4 5 6 7 8 9 10

Contents

CONTENTS

Acknowledgments

The following photos were reproduced with kind permission:

- Figure 1-1: Little Machine Shop, Pasadena CA
- Figures 1-2, 1-3, 1-4 and 1-8: Quality Machine tools (Precision Matthews), Coraopolis, PA

Special thanks go to the following, who provided "above and beyond" help and advice: Brad Bacon, for years my go-to source for sensible answers to (literally) hundreds of machining questions, and Ed Bindon and Chris Cleal, who were kind enough to read and clarify many sections of the text.

I am also thankful for the help and encouragement from Judy Bass (Industrial Press) and her colleagues, especially Judy Duguid and Patricia Wallenburg.

Richard Rex
Hendersonville, NC

Note from the Author

"Things I Wish Someone Had Told Me" is the title of just one chapter, but it really applies more generally to the entire book, which owes its existence to a host of patient instructors, mountains of reference materials, and—above all—to personal experience in problemsolving. I hope you will find it useful in forestalling some of the frustrating (and sometimes expensive) issues that crop up with any machine tool, and which certainly happened to me over the years.

INCHES TO MILLIMETERS CONVERSION

Most machine tools today come from countries using the metric system. Additionally, many projects call for metric measurements and metric hardware. To convert from inches to metric, and vice versa, all you need remember is the number **25** as a substitute for the "real" 25.4.

| inches to mm |

Multiply by **100**, then **divide** the result by **4**   **x 100 ÷ 4**
Example: 1" is approximately 100/4 = 25 mm

| mm to inches |

Multiply by **4**, then **divide** the result by **100**   **x 4 ÷ 100**
Example: 8 mm is approximately 32/100 = 0.32

In this book the word "mil" is sometimes used to signify 1/1000" (one-thousandth of an inch).

CHOOSING A LATHE

CONTENTS AT A GLANCE

RUBBER MATS CAN AVOID DISASTERS

Rubber mats around benches and machine tools can save the day when you accidentally drop something fragile, delicate, or sharp—think dial indicators, calipers, milling cutters. (The mats also provide foot relief.) Look for nonslip material, ideally 3/8" to 1/2" thick.

A SELECTION OF METAL SHOP LATHES

In some cases, machines similar in appearance to those discussed here are available from more than one supplier, but their overall quality and specifications may differ.

Mini Lathes

Mini lathes are in the range of 6" x 12" up to 7" x 16" (Figure 1-1), with outliers at both ends. Most of them have continuously variable spindle speeds and are capable of cutting screw threads, US and metric. The package usually includes a chuck, either 3 jaw or 4 jaw (most users will need both).

FIGURE 1-1   7" x 16" lathe with variable speed from 50 to 2500 rpm. Hollow spindle (3/4" clear). Supplied with a 3-jaw chuck and digital readouts on the cross-slide, compound, and tailstock. Includes a QC (quick change) tool post. Weight about 100 lbs.

There is also a size known as "micro," 4" x 6", which has nonstandard features and more limited capabilities (e.g., no screw cutting).

Larger-Model Shop Lathes

This class includes swing sizes of 10", 12", and 14" (for size definitions, see page 4). The largest size that can be considered "tabletop" is the 10", but most 10" users prefer a dedicated stand for stability. Specifications for larger lathes vary widely, starting with ac power input. Most 10" machines, plus a few 12" models, operate on 110-V single phase (Figure 1-2). Larger machines usually call for 220-V single phase (Figure 1-3), some three phase. Accessories included also vary across the board. Expect to see a powered cross-slide, faceplate, one or more chucks, and steady rests.

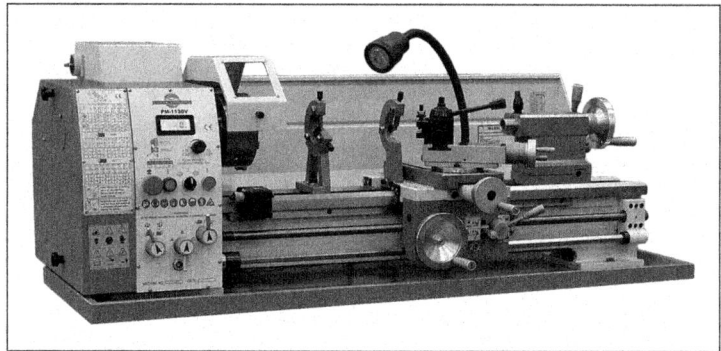

FIGURE 1-2 10" x 30" lathe with variable speed from 50 to 2000 rpm. Hollow spindle (1" clear). Often included are a faceplate, chucks, steady rests, a QC tool post, and a work light. Weight about 400 lbs.

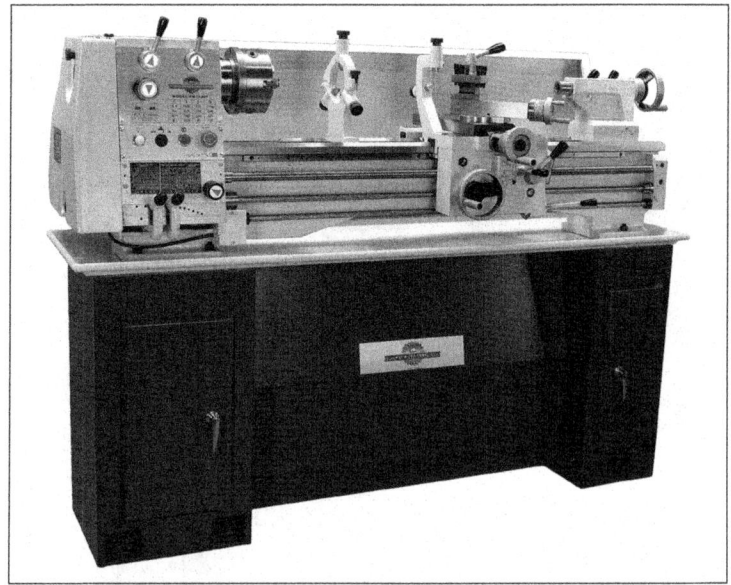

FIGURE 1-3   14" x 40" lathe, 220-V, single phase (2 hp) or three phase (3 hp). Traveling motor control switch on saddle. Gearbox-selected speeds from 50 to 2000 rpm. Hollow spindle (2" clear). Usually included are a faceplate, chucks, and steady rests. Weight about 1750 lbs., stand included.

1-1 LATHE SIZE

Example: 12" x 36" (see Figure 1-4)

The first number, 12, is the largest diameter in inches of material that can be "swung over the bed," meaning a cylinder of that size can be installed with its centerline on the lathe's centerline. This number takes no account of the height taken up by the saddle and cross-slide, so it doesn't mean you can machine a workpiece of that diameter in the ordinary way. The only way to machine something of the nominal 12" diameter is to attach it to a faceplate, clear of the cross-slide.

On larger "gap bed" machines, a 6" or 7" section of the bed just to the right of the headstock is cut away to allow extra swing close to the face-plate—an additional 5" diameter or more. When not in use, the gap is filled by an insert (Figure 1-5).

FIGURE 1-4 12" x 36" lathe, 2-hp motor, 220-V single or three phase. Traveling motor control switch on saddle. Gearbox-selected speeds from 90 to 1800 rpm. Hollow spindle (1-1/2" clear). Usually included are a faceplate, chucks, and steady rests. Weight about 1,000 lbs., stand included.

Another number to be concerned about is the diameter that can be swung over the cross-slide. This is a lot less than the lathe's nominal size and is typically about 6" or 7" for a 12" lathe.

The second number, 36, is the maximum length of the workpiece that can be installed between the headstock

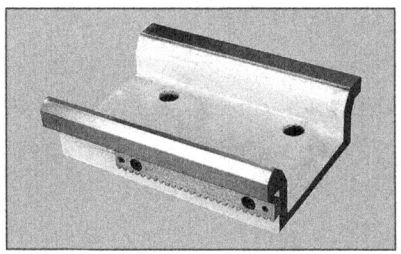

FIGURE 1-5 Gap insert.

(spindle) and tailstock centers. "Centers" are 60° cone-pointed steel cylinders ground to fit the tapered bores of the spindle and tailstock. The tapers vary from machine to machine: 7" and 9" lathes will likely have a Morse taper #3 (MT3) spindle and MT2 tailstock. Larger machines, 12" and up, usually come with an MT5 spindle and MT3 tailstock. If you are considering upgrading to a larger machine and have smaller MT accessories on hand, adapter sleeves are easily available.

1-2 HOW BIG A LATHE DO YOU NEED?

If you plan on miniature model making, you can probably get by with a 6" or 7" mini lathe. The next step up starts at about 10", a size that can handle a huge range of practical projects, including toolmaking and repair, auto and machine parts, and experimental work for the science lab. Machines 12" and larger are thought to be industrial, although there are many thousands of 12" and 14" lathes in home shops. Mini lathes and industrial-size lathes are all similar in principle but very different in capability. The following sections summarize the main differences.

1-3 WEIGHT

The heavier the better applies to all machine tools. For freedom from vibration and resistance to flexing under heavy cutting loads, there is no substitute for a mass of cast iron.

1-4 BENCH OR STAND MOUNTING

Most lathes larger than (say) 10" x 22" weigh 400 lbs. or more, calling for a reinforced bench or a dedicated stand. Stands with built-in cabinets are available for most lathes. A big advantage of the stand versus bench is the provision of bottom flanges for bolting down and/or leveling feet—if there's a choice, go for the type of self-expanding foot that can be adjusted from the top.

1-5 HANDLING THE LATHE

Except for the smallest machines, you will need an engine hoist or forklift to move the lathe around the shop. Almost invariably, larger lathes are shipped in two packing cases, one for the lathe and one for the optional stand. It is often convenient to install the lathe on its stand before moving it into position in one piece, as shown in Figure 1-6.

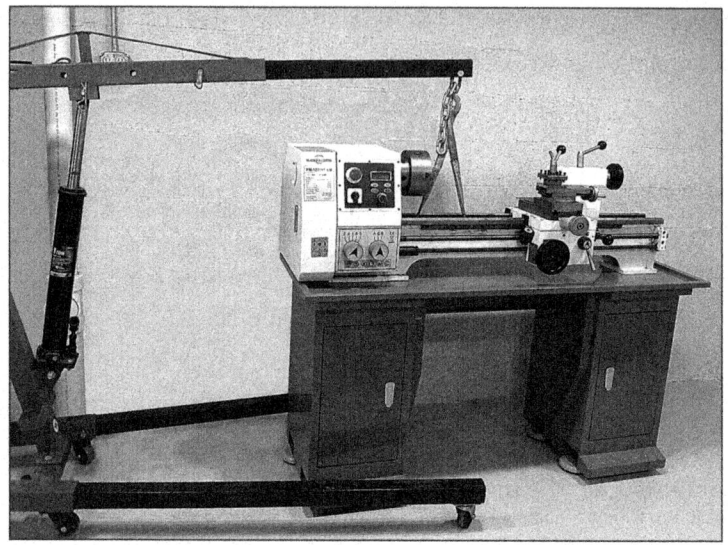

FIGURE 1-6 Lifting with slings. In this setup, the sling was looped through holes in the lathe bed casting. If instead the sling is set across and under the lathe bed, use spreaders to keep the sling clear of all components—especially the front shafts.

When selecting a location for the lathe, allow sufficient room at the back of the machine to allow access for installation of DRO (digital readout) scales, etc., and on the right side to allow removal/servicing of the lead screw and feed shaft.

Be sure to keep all lifting gear clear of any part of the lathe, especially the shafts at the front. If necessary, keep the sling away from the shafts using "2-by" spreaders.

Leveling is an important part of the installation—see Chapter 4.

1-6 SPINDLE BEARINGS

In a modern machine, expect to find tapered roller bearings or angular contact (deep-groove) ball bearings. Some call for regular lubrication; others are pre-greased and can be left alone for months, even years. No matter what its bearing system, the spindle must rotate freely—no notchiness—without a hint of radial or axial play. In other words, the chuck, or whatever is attached to the spindle, must feel rock solid.

1-7 SPINDLE BORE

Most midsize lathes, not all, have a hollow spindle with a bore diameter of 1" or more (Figure 1-7). This allows you to hold in the chuck long pieces of the commonly used bar sizes—longer even than the spindle itself—but watch out for the stock flailing around dangerously where it protrudes at left. This is correctable to some extent if the open end of the spindle is fitted with a "spider" attachment (a collar with three clamping screws at 120°).

FIGURE 1-7   Typical hollow spindle. This example, on a 12" lathe, has an ID of 1-1/2".

1-8 SPINDLE SPEED

Model shop lathes typically offer a range of six or more spindle speeds from about 50 to 2000 rpm. If the motor is single-phase ac powered, it has just one shaft speed. The various spindle speeds are derived from combinations of various pulleys and V-belts and an oil-filled gearbox (Figure 1-8). Of the two, a gearbox is the more convenient.

Even more convenient is a continuously variable speed from either a dc motor or a 3-phase ac motor controlled by a VFD (variable frequency drive)—see the next section, "Power Requirement."

1-9 POWER REQUIREMENT

Model shop lathes usually call for 110-V or 220-V single-phase ac. If you see a used lathe advertised with a 3-phase motor, this could mean that

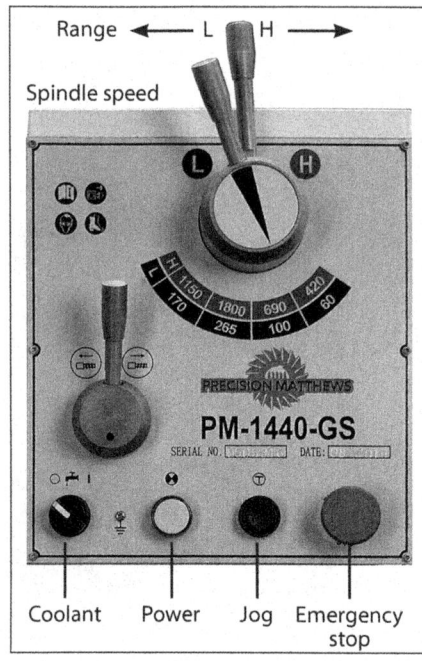

FIGURE 1-8   Spindle controls on a 14" lathe. "Jog," a feature seen only on larger lathes, is used to briefly nudge the spindle a few degrees around with each push of the button (forward or reverse usually selectable).

the shop needs to be wired for 3-phase (usually unthinkable), or it has a 3-phase motor running from a VFD. This is often done to achieve a wide range of continuously variable spindle speeds without the need for step pulleys or a gearbox. If you have a fixed-speed single-phase machine and would like greater flexibility, it can be a simple matter to replace the single-phase motor with a 3-phase motor of the same frame size, then attach an off-the-shelf VFD and potentiometer for speed control. Total cost can be as low as $300 for both items—see various forums.

1-10 CROSS-SLIDE AND COMPOUND DIALS

Many lathes sold in the United States, certainly the larger ones, have 10 TPI (threads per inch) lead screws on the cross-slide and compound. This means that one revolution of the handwheel gives a linear motion of 0.1". The dials are usually graduated in 0.001" intervals. Some have a second collar graduated in 0.2-mm intervals, a nice feature but not essential (Figure 1-9).

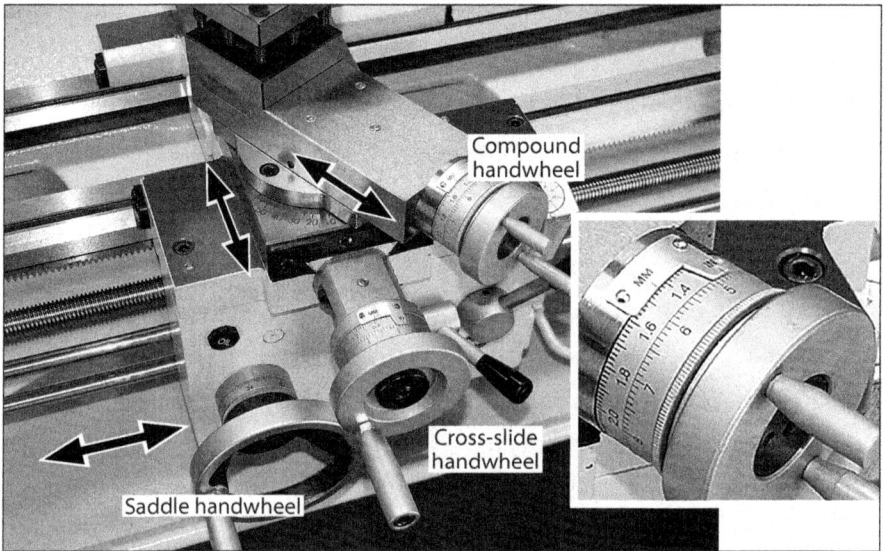

FIGURE 1-9 Cross-slide and compound. Both collars have standard (0.001") and metric (0.2 mm) graduations, an unusual feature.

Watch out for dials with an odd-looking total of divisions per revolution. *One example:* 60 divisions, claiming to be 0.001" per division. Not so. They are more likely to be 0.025 mm—close to 0.001" but not close enough for dead reckoning by counting divisions (a 1.6% error, in fact). This is a problem that goes away if a digital readout system is installed on the saddle and cross-slide (Section 1-30).

1-11 CHUCKS

The absolute minimum requirement for holding work is an independent 4-jaw chuck, "independent" meaning that the jaws are individually adjustable. They can also be flipped end for end to increase capacity. The 4 jaw can hold odd-shaped nonsymmetrical material (Figure 1-10) and

FIGURE 1-10 Odd-shaped workpiece in a 4-jaw chuck.

also be adjusted precisely for concentric running with round bar stock.

The downside of the 4 jaw is that it can take several iterations to set a round bar running true with a TIR (total indicator reading) of, say, 0.001".

TOTAL INDICATOR READING

TIR is the difference between max and min readings on a dial indicator in contact with the rotating workpiece. If the workpiece is a perfect cylinder, it may be possible to adjust for less than 0.001", but the fact is that most bar stock is not truly round—checking with calipers at various points around the circumference can be an eye-opener.

For round stock, a less time-consuming solution is the self-centering 3-jaw chuck (Figure 1-11). This is the workpiece holder of choice in most small engineering shops, so much so that in many cases it is almost a permanent fixture on the lathe. A very useful feature of the 3 jaw is its ability to hold hexagon-section material, something a 4 jaw simply cannot do.

Self-centering chucks are operated by an internal scroll (Figure 1-12). At the outer rim of the scroll's back surface is a bevel gear that is turned by any of three bevel-gear stub axles, indicated by arrows in Figure 1-12. All three internal gears move together when any one of them is turned using the square-tip wrench (but it's a good

FIGURE 1-11 Typical 3-jaw chuck. Before doing anything with a chuck, or faceplate, protect the bed with scrap wood. The arrow points to one of three adjusters.

idea to check for consistent tightness on all three before running the lathe). Because their jaws cannot be flipped end for end like those on an independent chuck, self-centering chucks come with two sets of jaws, internal for smaller workpieces and external for larger.

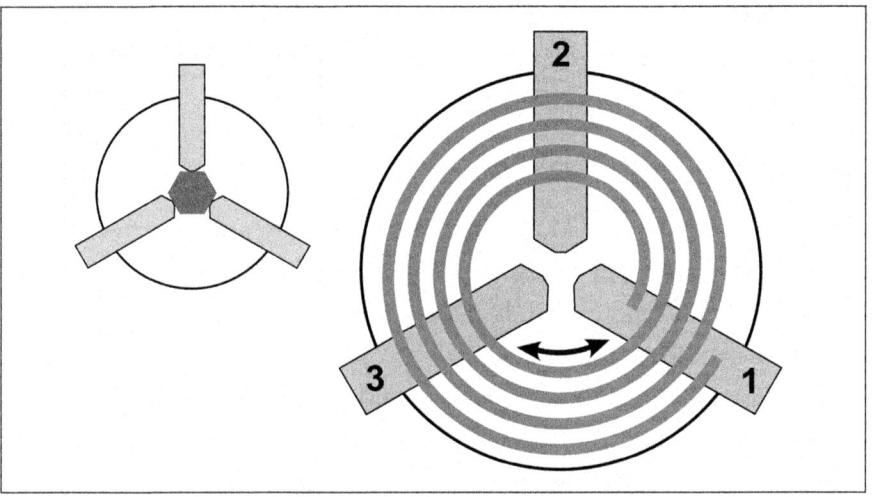

FIGURE 1.12   The 3-jaw chuck scroll. The 3-jaw chuck works with both round and hexagonal bar stock.

The remarkable fact about self-centering chucks is that most of them—even the budget variety—can hold a circular or hexagonal workpiece with reasonably small runout, often as little as a few thousandths of an inch. Don't expect much better than that, unless you have a higher-class chuck with adjustable jaws. Also be aware that holding accuracy can vary with workpiece diameter. This is mostly due to inconsistencies in the scroll.

HOLDING HEXAGONAL STOCK

This is a capability of the 3-jaw chuck that can be a great bonus when machining custom screws. Die-cutting screw threads puts a lot of slipping torque on the workpiece, often to the point where damaging force is used to tighten the chuck. Hexagonal bar can be firmly held with very little strain on the chuck.

An interesting side note: In place of a scroll, a 4-jaw independent chuck has four adjusting screws at 90° intervals. The jaws themselves are quite different, too: Their "inner teeth" that engage the screws are not arc-shaped (as they are in a 3-jaw). Instead, they are straight (but angled), which allows the jaws to be inserted either way around, in the same way that a nut can be flipped 180° and still run onto a screw.

1-12 WHAT SIZE OF CHUCK?

This is a personal choice. Most lathes up to 12" come with 5" or 6" diameter chucks, 3-jaw and 4-jaw. While 8" chucks may be offered for 12" and larger lathes, they can be quite a handful (weight about 25 lbs.), calling for a rugged spindle and bearings to cope with the load.

Because they are so much easier to handle, after-market 4" or 5" chucks are the better choice in many applications. They may not be available off the shelf for your lathe, but there's an easy fix for that—buy a backplate to match the spindle; then machine it to suit the chuck.

1-13 FACEPLATES

Faceplates are usually just a little smaller than allowed by the lathe's "swing"; for example, a 12" lathe may be shipped with a 10" faceplate (Figure 1-13). Hardware to attach a workpiece to the faceplate is not supplied and is up to the user. When attaching a workpiece to the faceplate, it can sometimes be helpful to do the initial setup on the

FIGURE 1-13 D1-5 faceplate.

bench and then to fine-tune it with the faceplate reinstalled. If the workpiece is seriously asymmetrical, install balancing weights to lessen wobble.

1-14 HOW DO CHUCKS/FACEPLATES ATTACH TO SPINDLES?

There are several attachment methods, each with its own pros and cons:

1. Threaded spindle nose, mating with internal threads in the chuck backplate. **Upside:** Easy to install if the chuck is light-weight. **Downside:** Tends to unscrew if the spindle is reversed unless secured by safety clips, etc. (not all of them are).
2. Threaded studs on the chuck backplate fit into three holes on the spindle flange, secured by nuts on the headstock side of the flange. **Upside:** Secure in both spindle directions. **Downside:** Can be awkward to install and tighten the nuts.
3. Type D1 camlocks, most often seen on large lathes, 12" and up, and occasionally on smaller ones. The most commonly used camlocks in the small shop are D1-4 with three studs and D1-5 with six studs (don't look for logic here). D1-4 studs are 5/8" diameter; D1-5 studs are 3/4".

Taking D1-4 as an example (Figure 1-14), the three studs are threaded into tapped holes on the chuck backplate. Each stud has a D-shape crosscut to engage a corresponding cam within the spindle nose (Figure 1-15). The function of the cams is to pull the chuck back-plate inward to locate its internal taper firmly on the spindle nose.

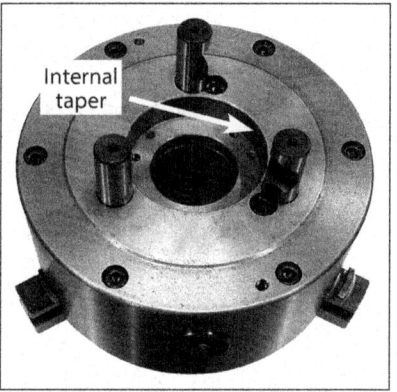

Internal taper

FIGURE 1-14 D1-4 chuck.

To install the chuck, set the cam markers on the spindle to 12 o'clock; then push the chuck home (Figure 1-16). While supporting the full weight of the chuck, turn the cams clock-wise to locate them snugly. Before fully tightening them, check that each of the cam markers lies between 3 and 6 o'clock, marked by two Vees stamped on the spindle (Figure 1-17).

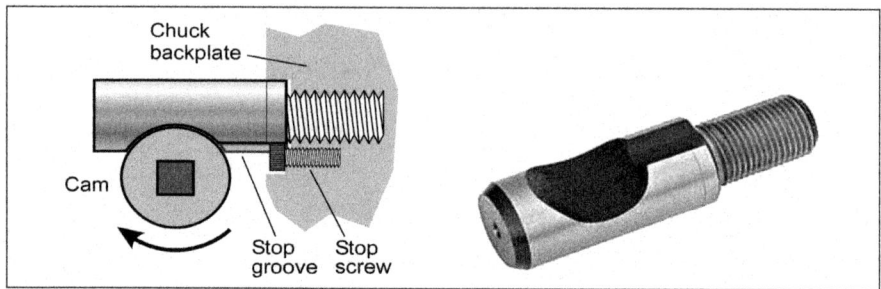

FIGURE 1-15 Camlock stud. The stop screw doesn't clamp the stud in place. It prevents the stud from unscrewing accidentally.

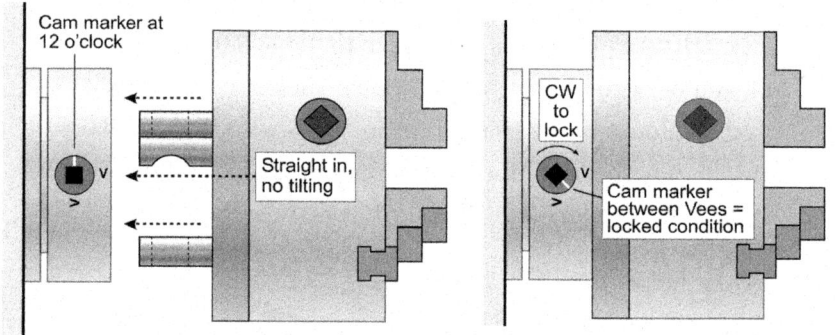

FIGURE 1-16 Installing a camlock chuck.   **FIGURE 1-17** Cam in locked condition.

The upside of camlocks is easy, reliable installation, provided the cam markers are properly positioned. They have the obvious downside that running the lathe with misaligned markers is dangerous (imagine a weighty chunk of cast iron wobbling off the spindle).

1-15 STEADY REST AND FOLLOWER REST

Most lathes come with a steady rest (aka "center rest") and follower rest similar to those in Figure 1-18. They are used to stabilize *long cylindrical workpieces* that would otherwise be deflected by the cutting tool.

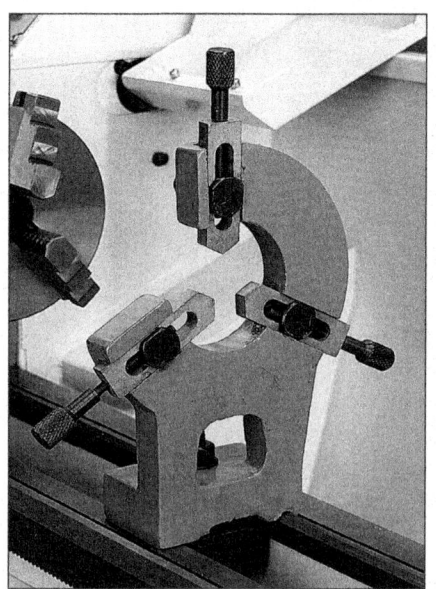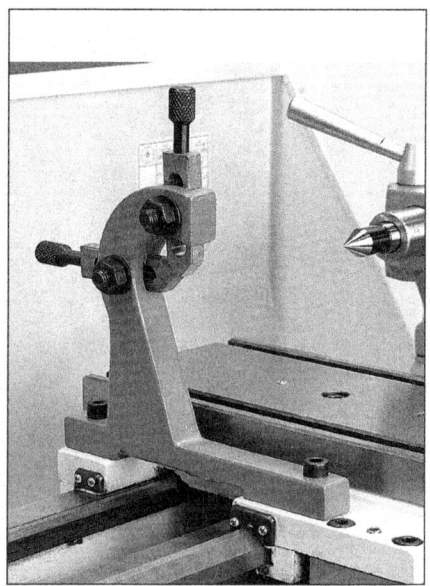

FIGURE 1-18   Steady rest (*left*) and follower rest (*right*).

The steady rest is secured rigidly to the lathe bed between the saddle and chuck. It is positioned to minimize the distance between it and the cutting action. The follower (aka traveler rest) is quite different: It is attached to the lathe saddle, just to the left of the cutting tool. The follower jaws effectively eliminate upward and backward deflection of the workpiece. The follower is used either on its own or in combination with a steady rest (Figure 1-19).

Setting up the steady rest is rarely straightforward. The surface of the workpiece to be stabilized by the rest has to be (1) truly round, and (2) truly on the lathe's centerline. (The three jaws of the steady don't help—they're not calibrated, and they're completely independent.) If the workpiece, assumed to be "long and thin," can be contained within the spindle, with only an inch or two protruding from the chuck jaws, the starting point could be to skim-cut its outer end. The rest would then be *temporarily installed* so that its jaws could be closed on the workpiece with even

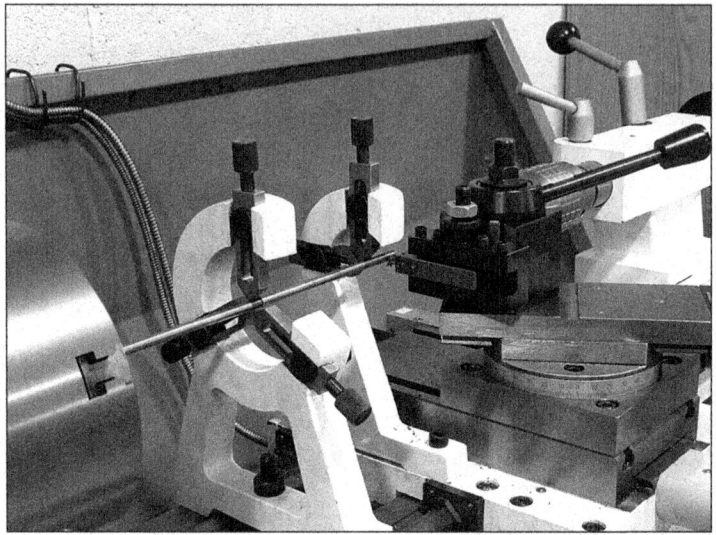

FIGURE 1-19   Screw cutting with steady and follower rests.

pressure (and with no danger of pushing it off-center). The steady rest can now be reinstalled in its working location, with the workpiece repositioned accordingly. The jaws of the follower rest are adjusted in a similar way. Be sure to lubricate the jaws.

1-16 LATHE CENTERS

Most lathes are supplied with a pair of hardened centers with tapered shanks, one to fit the spindle's internal taper and one for the smaller internal taper in the tailstock barrel. The tapers are usually Morse taper (MT) numbers. For a typical 10" lathe, expect the numbers to be MT4 spindle and MT2 tailstock; for a 12" lathe, MT5 and MT3.

A workpiece to be mounted between centers is drilled at both ends with a center drill, as shown in Figure 1-20 (see also Section 2-27). The drill leaves a cone-shaped 60° cavity with a small divot at the bottom to clear the lathe center's point.

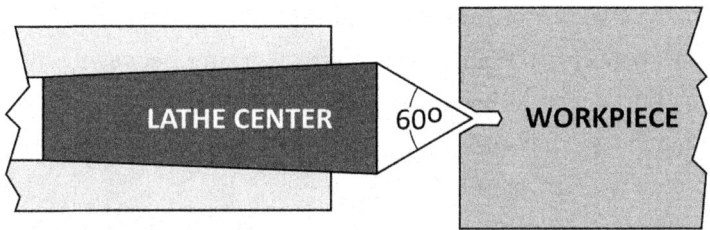

FIGURE 1-20   Lathe center. The outfacing point on all centers, regardless of size, has an included angle of 60°.

The tailstock center is used mostly to support the "far end" of long workpieces, sometimes with additional support from a steady rest or follower rest. The near end of the workpiece may be held in a 3-jaw or 4-jaw chuck. Alternatively, for better accuracy, the near end may be held by the spindle center and driven by a "lathe dog" (Figure 1-21).

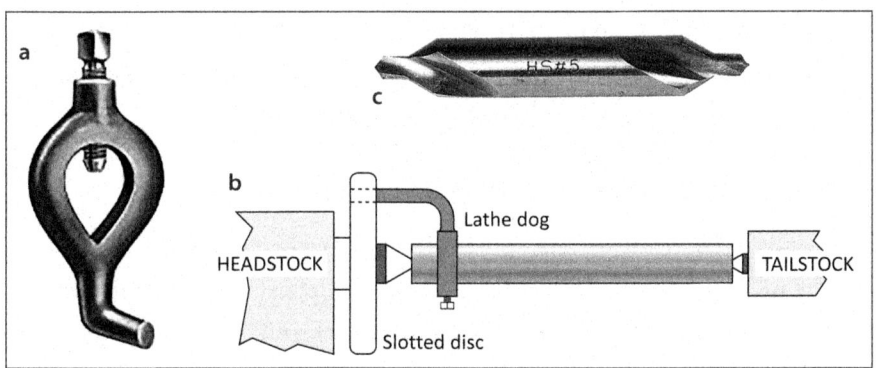

FIGURE 1-21   (a) Lathe dog. (b) Turning a workpiece between centers. (c) Center drill.

At one time, many years ago, most lathes came with a radially slotted disc that mounted on the spindle nose, the slot being the means of imparting drive to the workpiece. Lathe dogs are still available, but to drive them you will need to improvise, maybe using a faceplate as noted in Section 4-7.

At this point, you may be asking yourself this question: *Why "turn between centers" when a chuck is a perfectly good means of driving the workpiece?*

There are two reasons: One is *accuracy!* Because the spindle center is likely more concentric than a chuck; also working between centers, you may be able to cut to a midpoint, flip the workpiece end for end, and then cut the other end to blend perfectly. The other reason to turn between centers is *cutting a long taper* by offsetting the tailstock. (The downside of this is the need to spend a fair amount of time resetting it later for parallel turning.)

If you plan on turning between centers regularly, you may want to invest in a live center, the point of which rotates with the work (Figure 1-22). If using the standard static cen-

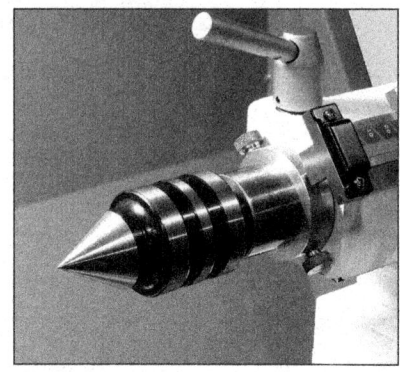

FIGURE 1-22 Live center in the tailstock.

ter, lubricate it often. In either case, watch out for overheating from the cutting action. This can cause a long, thin shaft to buckle.

1-17 OTHER WORKPIECE HOLDING ACCESSORIES

Professional machine shops sometimes prefer to use collets instead of chucks to hold bar stock and cutting tools. Collets also have their place in the model shop, most often in milling machines, but there are several factors that make them less desirable for the model shop lathe.

First, a few words on what collets do that sets them apart. The two types of collet most often used on lathes are 5C and ER (Figure 1-23), which have been around for many years (the 5C style from Hardinge, made in the United States, and the ER style from Rego-Fix, made in Switzerland). Both types are closed by screw action that pulls them into an internally tapered collet chuck attached to the machine spindle.

ER and 5C are both open-back designs, meaning that they can hold smooth bar stock and tool shanks of any length (assuming "reasonable" overhang at either end of the collet) with good concentricity and repeat-

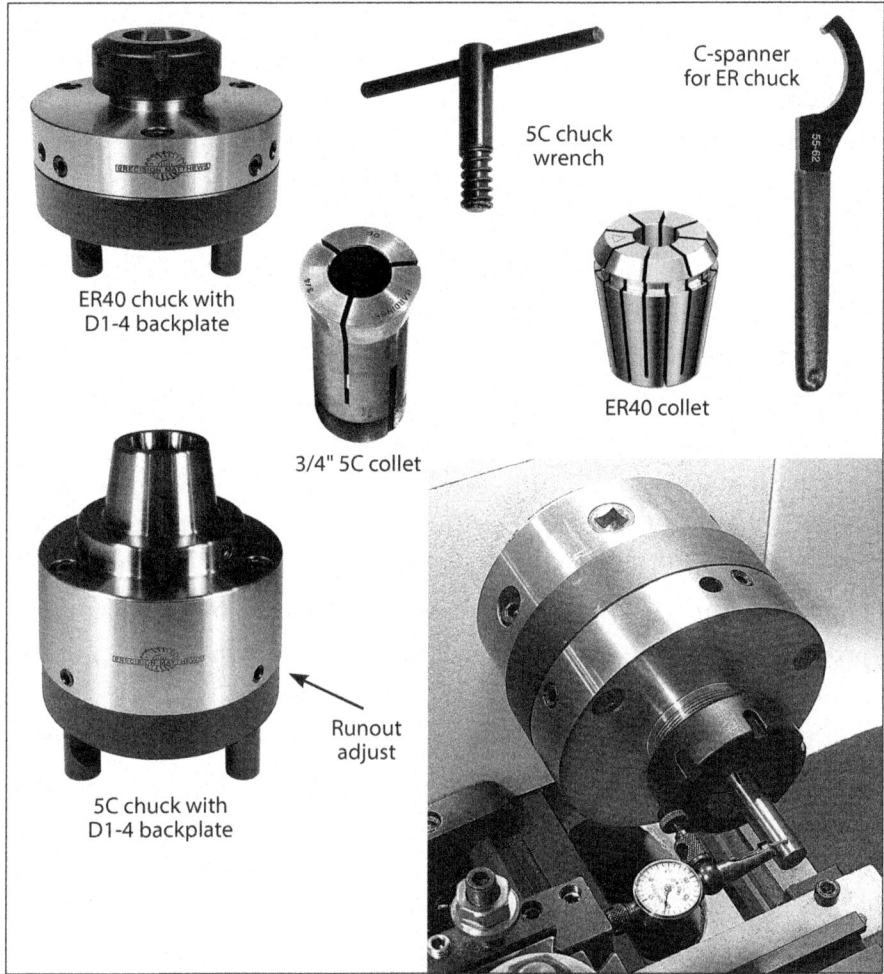

FIGURE 1-23 ER and 5C collet chucks. Both ER and 5C collets require a special internally tapered chuck. The ER collet is slit at both the inner and outer faces to give a tighter, more uniform grip. Several versions of the 5C collet are available for various types of bar stock, including square and hexagonal sections. The collet chucks shown here can be adjusted for precise concentricity. Check using ground stock, as shown here.

ability. The downside of both is their limited range of closure, not much more than 0.02" for 5C, 0.04" for ER. This means you need a collet for each of the nominal stock or tool shank diameters you plan to use. 5C collets, but not ERs, are available for square and hexagonal sections.

Unlike 5C collets, which all *have the same outer dimensions*, ER collets—and their matching chucks—come in eight sizes, from ER-8 to ER-50 (the number is approximately the outer diameter of the collet nose, in mm). Another difference, compared with 5C, is that the ER collet closes uniformly at both ends, thus gripping more reliably with less tendency to wobble.

Several styles of collet chuck are available for the larger-model shop lathes, usually with camlock spindles. The better types of collet chuck, as shown in Figure 1-23, have four set screws on the perimeter that push on an internal shoulder on the front face of the chuck backplate. The screws allow semipermanent adjustment to minimize runout—usually a one-time procedure, unless one particular collet itself is out of true.

The 5C chuck draws in the threaded neck of the collet by rotating an internally threaded bevel gear. This is in turn rotated by an external T-wrench similar to a standard chuck key. The ER chuck, a much simpler design, has a threaded nose and a ribbed nut tightened by a C-spanner. A popular chuck size, ER-40, accepts all ER-40 collets with IDs from under 1/8" to 1".

The strong suit of collets, compared with the typical self-centering chuck, is their ability to grip the stock or tool tightly and concentrically within a mil or so, at least in theory. This is true in practice, too, provided you have industrial-grade collets and a collet chuck of matching quality to go with them. High cost is the reason they may not belong in a budget operation. You can buy import 5Cs and ERs for a few dollars apiece, but their performance is unpredictable.

1-18 TAPER TURNING

Taper turning makes most model shop machinists a little nervous, because the usual objective is to *match precisely* the taper of a mating component, such as a tailstock bore. It is impossible to do this simply by referring to graduated scales; instead, use an existing taper as a template (Figure 1-24). Dedicated taper turning attachments are rare and expensive (Figure 1-25). If an attachment is not available for your lathe, there are two other methods: (1) For *long tapers*, offset the tailstock, as shown in Figure 1-26. (2) For *short tapers*, set a specific angle on the compound slide (but this method is limited to the travel available on the compound, usually about 3-1/2").

FIGURE 1-24   Setting the compound angle for a tapered tool shank.

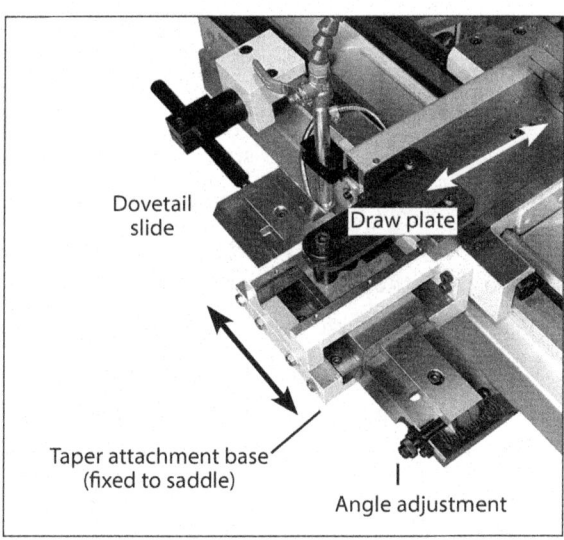

FIGURE 1-25   Dedicated taper turning attachment. Not available for all lathes. To use a taper turning attachment like this, the cross-slide is first uncoupled from its lead screw (making the handwheel inoperative). It is then coupled, through the draw plate, to the taper attachment.

Dovetail slide

Draw plate

Taper attachment base (fixed to saddle)

Angle adjustment

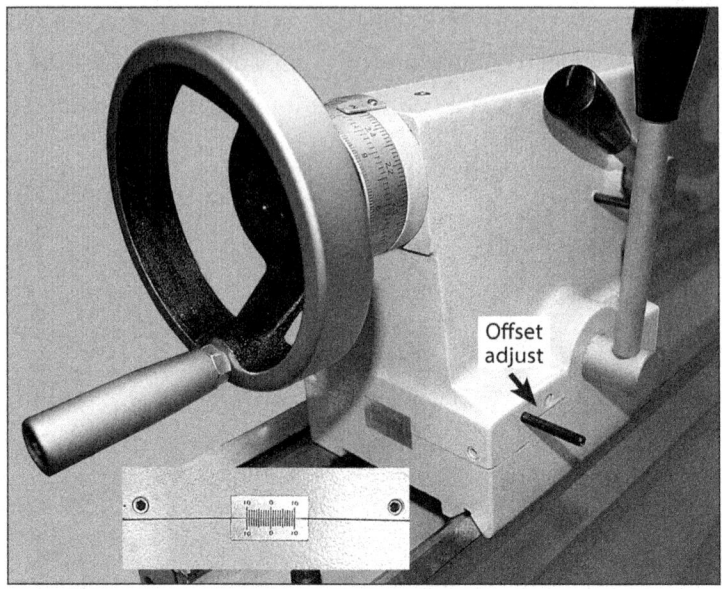

FIGURE 1-26   Tailstock offset. Most tailstocks have a graduated scale (inset).

1-19 SCREW CUTTING AND POWER FEEDING

To cut screw threads on a lathe, the saddle is driven by a lead screw, which is in turn powered via a gear train from the spindle. (The number of threads per inch is determined by the ratio of spindle speed to lead-screw speed.) Screw cutting is just one specific use of power feeding. More routinely, power feeding is used to automate routine turning with a knife tool, often delivering better surface finish than is possible with hand feeding.

Larger lathes have separate systems for the two functions; they also typically power-drive their cross-slides, which is rarely the case with budget lathes. In basic lathes, the *lead screw* may be the *one and only* means of power feeding. This is a workable arrangement for light-duty machines, but in some cases it may deliver feed rates too high for regular turning (see key fact #1 in Section 1-24).

The gearbox in more capable machines has a second output shaft, the *feed shaft*, dedicated to routine turning and facing (as opposed to screw cutting), as shown in Figure 1-27. In some mid-size lathes, a full-length keyway on the lead screw provides virtual second-shaft capability, delivering a full set of lower feed rates for routine work (without wearing the lead screw threads).

In 12x and 14x lathes, there is often a third full-width shaft below the lead screw and feed shafts. This is the motor control shaft (usually controlled by a lever) that allows the operator to select Forward-Stop-Reverse without taking eyes off the work (Figure 1-28).

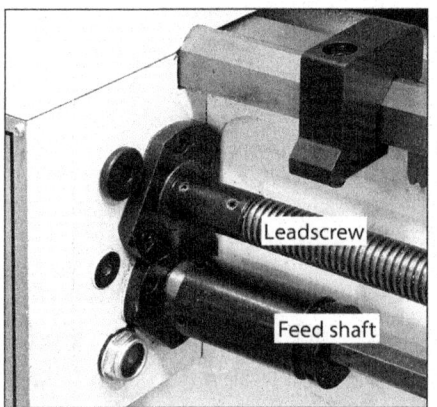

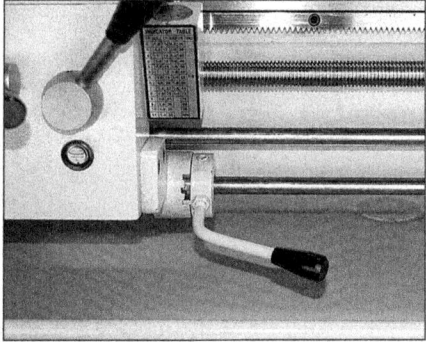

FIGURE 1-28 Typical motor control lever.

FIGURE 1-27 Separate lead screw and feed shaft.

1-20 GEARBOX OR EXTERNAL GEARS?

Power feeding and screw cutting both call for selectable ratios between the spindle speed and the saddle/cross-slide feed mechanism. On most lathes, this is provided by a combination of external loose gears on the left side of the headstock and a gearbox that drives a lead screw for screw cutting (Figures 1-29 and 1-30). Basic lathes may have external gears only.

FIGURE 1-30   Representative power-feed gearbox. In this example, the right-hand knob selects feed shaft drive (X) or lead screw, for screw cutting W and Y).

FIGURE 1-29   Representative external gears.

1-21 EVERYDAY POWER FEEDING VERSUS SCREW CUTTING

All lathe operations have to do with metal removal. Most operations are familiar and straightforward—even obvious—such as face-cutting a workpiece or reducing its diameter by traversing a knife tool, both of which are everyday actions. They can, if you're fortunate, be eased by power feeding. Screw cutting is also metal removal, but in a very specific way. Screw cutting has so many variables to consider that it has a separate chapter—Chapter 5.

1-22 SADDLE CONTROLS

Power feeding is almost always controlled by levers on the saddle (Figure 1-31). The layout of the various functions varies widely, but all lathes with *screw cutting capability* have one common feature, namely a *split nut* that engages the lead screw only when its lever is down. When the lever is up, the nut disengages, leaving the saddle free to be moved by its handwheel or feed shaft (Figure 1-9).

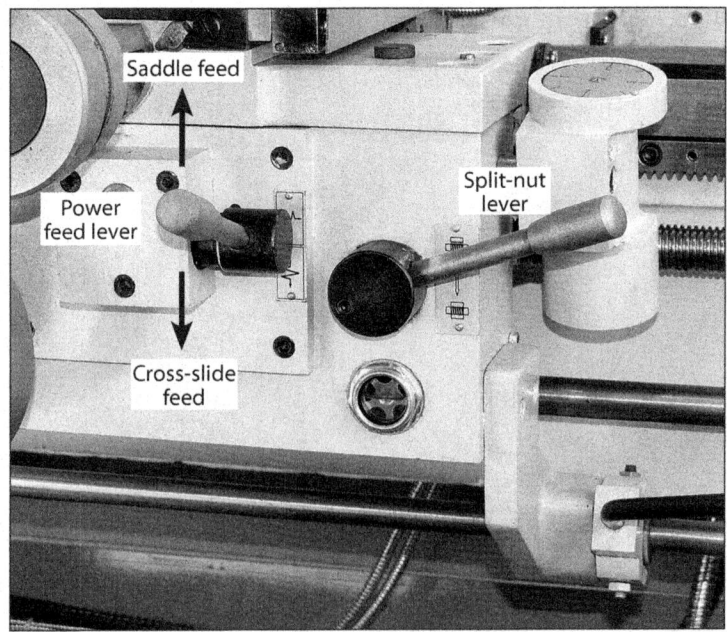

FIGURE 1-31 Typical power-feed and split-nut controls. On this lathe, which has both a feed shaft and a lead screw, the lead screw is used only for screw cutting. It is engaged by the split nut.

1-23 POWER-FEED DIRECTION AND LEFT-HAND THREADS

For everyday turning operations, when power-feeding, you can expect that the saddle will move to the left—toward the headstock—as the spindle revolves forward (counterclockwise, as viewed from the tailstock). Likewise, the cross-slide, if powered, can be expected to feed inward, toward the lathe axis. There are times when it is necessary to have the saddle and cross-slide move in the other direction, so there needs to be a means of reversing the entire power-feed system.

If there is no such reversing capability, it may be possible to install an idler in the external gear train, as shown in Figure 1-32. This is what I had to do on an earlier 10" x 22" lathe to cut left-hand threads. The idler I used in this case was one of the supplied change gears for which there was no obvious need. It could be swung into and out of mesh in just a few seconds and locked in either position.

FIGURE 1-32   Reversing idler gear on a shop-made bracket (shown disengaged).

1-24 POWER-FEED KEY FACTS

1. For regular turning operations the saddle feed rate sometimes needs to be as slow as 0.002" per revolution of the workpiece.
2. For facing operations the powered cross-slide feed rate is usually about one-half of the saddle feed rate.
3. The saddle feed rate for regular turning is a lot slower than for *any screw cutting*, even for threads as fine as 80 TPI, which is 0.0125" per revolution of the workpiece.

1-25 THE MAIN POWER-FEED QUESTION TO ASK

We'll assume that your chosen lathe offers all the inch/metric thread pitches you will need, together with a reversible lead screw (see below) and a usable range of power-feed rates—so far so good; but you also need to know that switching from one thread pitch or feed rate to another will be *acceptably* quick and easy. Is it simply a matter of turning shifter knobs on the gearbox? Or is it a combination of turning knobs *and* swapping external gears? So, if screw cutting is important to you, the question really comes down to how much effort it will take to go from one thread to another within the chosen system of measurement, US or metric.

Bear in mind that there is *always* gear swapping when switching from US to metric threads and vice versa. This has to do with adding or removing a *transposing gear*, and is something we all have to live with. If your lathe has a threads-per-inch lead screw, usually 8 TPI in the United States, you have to transpose to cut metric threads. Likewise, if you have a metric lead screw, a rare item in the United States, you will have to transpose to cut US threads.

One last issue to do with screw cutting: It is important to see if the lathe's lead screw and feed shaft can be *reversed relative to the spindle*. This should be possible on every lathe, but it's worth checking for the odd occasions when you need to cut a left-hand thread. Is it straightforward, or difficult to do, even calling for shop work, as in Figure 1-32? Ideally, there will be a control panel lever to select left or right saddle motion, as shown earlier in Figure 1-8.

1-26 LATHE REFINEMENTS

"Refinements" may be the wrong word here—for most users, these are absolutely essential, starting with the two most mundane items—backsplash and chip tray. The backsplash ceases to be mundane the first time you spray the machine shop wall with cutting oil. If it's an extra for your chosen lathe, just buy it, or rig up your own wall protection with hardboard.

For a chip tray, look for a heavy-gauge aluminum baking pan as large as will conveniently fit under the machine, up against the back wall of the lathe tray. The chip tray needs to be shallow enough that it can be slid forward without hitting the saddle handwheel.

One other add-on you will need, probably from day one, is a tailstock chuck. Also, it will not be long before you begin to think of DROs for the tailstock, saddle, and cross-slide, truly a game changer (Sections 1-29 and 1-30).

KEEPING THE SHOP TIDY

All machine tools generate a surprising amount of debris that can bring shopwork to a halt unless cleaned up regularly. In the lathe area, the problem is 90% controlled by the backsplash and by a chip tray that can be slid out from under the lathe bed. For the remaining 10%, consider disposable paintbrushes and a shop vac (Chapter 3).

1-27 TAILSTOCK CHUCK

This is one of the most frequently used lathe accessories. It is installed in the tailstock internal taper, which is usually MT2 or MT3, depending on lathe size. Many tailstock chucks are assembled from two pieces, the chuck itself and an MT shank, the other end of which has a stubby Jacobs taper to hold the chuck. This is a workable arrangement, but it isn't as sturdy or accurate as a one-piece precision chuck (Figure 1-33).

Most tailstock chucks today are keyless, which users seem to prefer (surprisingly, they grip drills, etc., almost as reliably as keyed chucks). Chuck capacity is important, not just at the upper end (usually 1/2" or 5/8"). Many chucks have a lower limit of 1/8", good only down to a #30 drill. A more useful (rare) lower limit is 1/32", good for the entire number

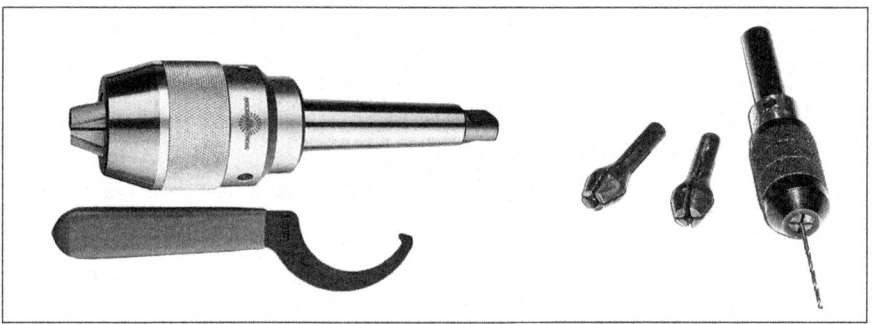

FIGURE 1-33 Typical one-piece precision chuck. *Right:* Micro drill chuck adapter with three collets for small drills. The drill bit shown here is #60, 0.040" diameter.

drill range. Yet more rare is a genuine zero to 1/2" chuck; these chucks are available, even if you have to go with a 2-piece Jacobs + MT shank. If your chuck is 1/8" minimum, for small drills you will likely need a micro drill chuck adapter from watchmaker or model engineer catalogs. Buy the best you can find—the cheaper ones usually run way out of true.

1-28 QCTP INSTEAD OF A 4-WAY TURRET

Most lathes are supplied with a 4-way turret toolholder that allows any of four cutting tools to be indexed quickly into position (Figure 1-34).

This sounds convenient, but it isn't. For one thing, the tools you installed yesterday are probably not the ones you need today; for another, there are some tools the 4-way turret can't easily accommodate—cut-off blades, boring tools, etc. But by far the most frustrating thing about the 4-way is the fact that each tool has to be individually shimmed to bring its cutting

FIGURE 1-34 A 4-way tool turret.

edge up to the lathe's centerline. If you sharpen the tool, shim again; ditto if you exchange the tool.

For all these reasons, most machinists will tell you the 4-way has to go. Replace it with a quick change tool post (QCTP), shown in Figure 1-35. Anyone out there still shimming lathe tools, *please note* the QCTP is a game changer.

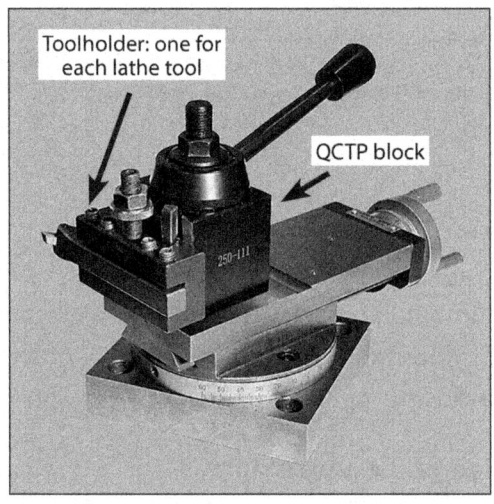

FIGURE 1-35   Aloris-style QCTP with an RH knife tool in the toolholder. This is a Series 100 QCTP for 10" and 12" lathes; 14" lathes usually need the Series 200 QCTP. The block has two locations for the toolholder, the left-facing one shown here, the other facing the back. In some situations, the locking lever can obscure the work area or be otherwise inconvenient. For that reason, on both of my QCTPs I have drilled and tapped an alternative lever location on the tapered upper collar.

The centerpiece of the QCTP is a heavy square-section block with dovetails on two sides for toolholders; the side dovetail is for everyday (external) cutting tools, and the rear dovetail is for boring tools. It can be *freely rotated about its vertical axis* and locked for any desired side-cutting angle—no need to modify the tool geometry. Most importantly, each tool has its own toolholder, which comes with a screw-adjusted height setting—set it once and forget it, a *priceless advantage*. If you sharpen it by grinding off a few thousandths, simply tweak the height adjustment, and you're done. No more shimming and *no need to check height when tool swapping. Just one caveat:* Bearing in mind the variety of tools most projects call for—knives, boring bars, cut-off blades, thread cutters, etc.—plan on acquiring a sizable collection of QCTP toolholders.

QCTPs come in two main styles, "piston" and "wedge." The wedge style is the more expensive and is said to deliver more repeatable tool placement than the piston. (I have one of each type and have not seen an appreciable difference.)

Exchanging the 4-way turret for a QCTP is usually a simple operation, in some cases a direct replacement. Whatever it takes, it's worth it.

1-29 TAILSTOCK DIGITAL READOUT (DRO)

This is not usually supplied, but it is fairly simple to install in the shop (Figure 1-36). It is a great help in determining the exact depth that a drill has entered the workpiece—simply *zero the DRO* when the nose of the drill is at the work face. Suitable DROs are available from several suppli-

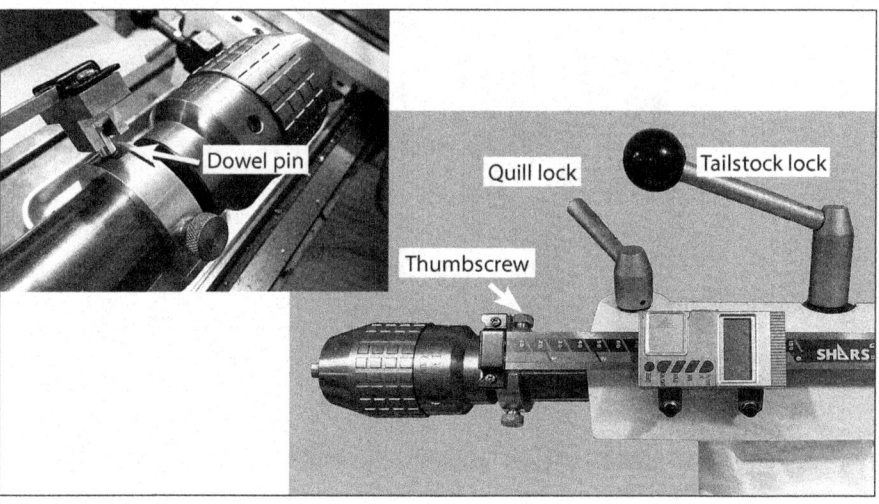

FIGURE 1-36 Tailstock DRO. A rotation-proof attachment method is shown at left. The dowel pin fits closely in a U-shaped block. Backlash is not a serious concern—we are usually interested only in the forward direction.

ers. These are not glass scales, so they can (usually) be cut to any length with a Dremel cut-off disc. Go very slowly; don't overheat. *One caveat:* All tailstock barrels rotate a degree or two, often more. This can't be allowed to displace the DRO scale. In the example shown, a shop-made collar is attached by thumbscrews to the barrel nose, and a dowel pin on the collar engages in a U-shaped fixture at the end of the scale.

1-30 SADDLE AND CROSS-SLIDE DRO

This is perhaps the most important option of all (Figure 1-37). The DRO most often used on today's lathes is a 2-axis system based on high-precision glass or magnetic scales. *Here's the thing* . . . in just a few minutes of hands-on experiments, you will see what a game changer the DRO can be, compared with yesterday's cut-stop-measure-cut operations.

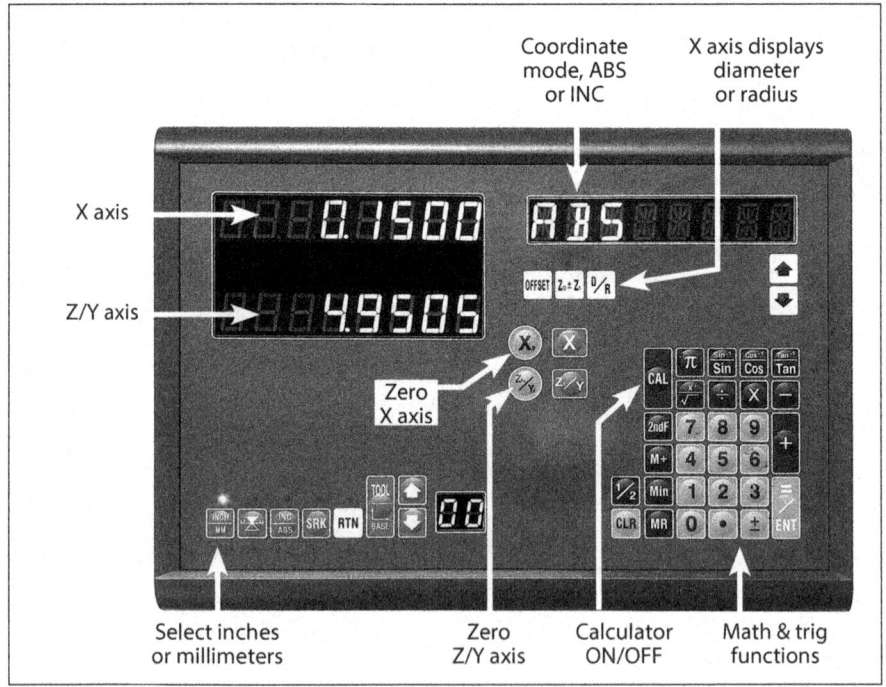

FIGURE 1-37 Typical 2-axis DRO display.

Suppose you need to reduce the diameter of a bar over a distance of 1.375" from the end face. With a DRO, you would do this by setting the tool tip gently against the outfacing end of the workpiece, then zeroing the Z display by pressing the Z_0/Y key, shown in Figure 1-37. (Note that some users call left-right saddle motion "Y"; others—more in line with general practice—call it "Z.") Having set the X axis (cross-slide) to the desired depth of cut, you then make a cutting pass from right to left, stopping when the Z/Y display is exactly −1.375. Successive cutting passes would be made in whatever decrements you choose, homing in on the target diameter in a completely predictable way. Just before making the finishing cut, you would typically zero the X display (X_0 key)—without in-feeding the tool—and then check the workpiece diameter. You can now feed in the cross-slide by the exact difference between the actual and desired diameters.

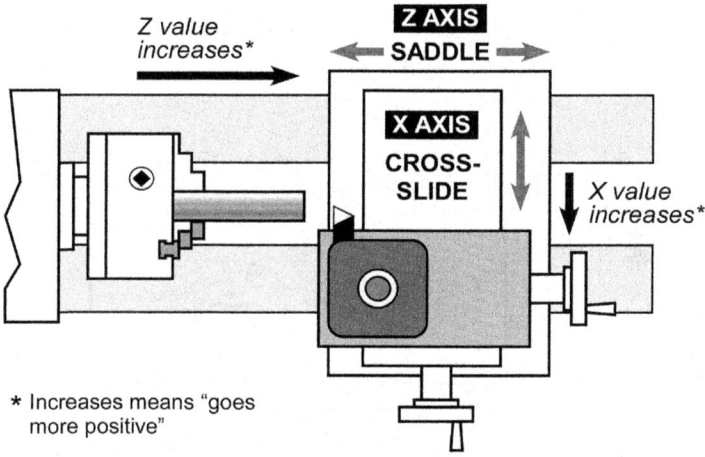

FIGURE 1-38 Typical lathe DRO setup. The "increasing value" directions shown here are arbitrary and can usually be reversed. The terms Z and X may differ on some DROs.

1-31 INSTALLING A DRO

DROs are not difficult to install, but careful alignment is essential, and that takes time. Some suppliers offer an installation service.

One question for the installer: With the cross-slide scale installed, will the following still be accessible: *cross-slide lock* screw, *cross-slide gib* screws, and *saddle lock* screw? Be concerned if you hear "That's not a problem; we install the cross-slide scale on the left." This is *not a good answer*, because it places the DRO close to the chuck (plus it calls for greater overhang of the cutting tool). It also puts the scale in the worst location for contamination by cutting fluid and chips.

There are several ways to install the scale on the right side and still have the access you need. Figure 1-39 shows one solution that takes almost no extra effort and doesn't seriously get in the way of saddle motion. It uses a couple of standoff bushings, and it substitutes hex-head lock screws in place of the regular set screws. The photo in Figure 1-40 is a close-up showing where the saddle lock is usually found.

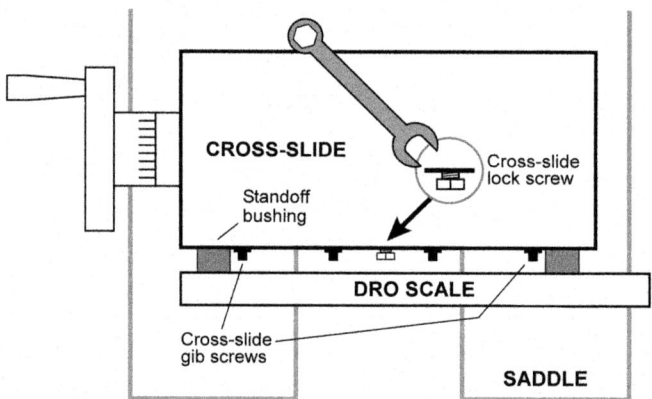

FIGURE 1-39 Installing the cross-slide scale. Standoffs allow access to the lock screw and gib screws and also the saddle lock, Figure 1-40.

FIGURE 1-40 Typical location of the saddle lock.

1-32 SADDLE STOP

This is often included as a standard accessory (Figure 1-41). If not, it is probably something you can get along without—or make in the shop.

The saddle stop can be clamped anywhere along the lathe bed. It is used *only when manual feeding*, to stop the saddle repeatedly at a specific point on the workpiece, maybe for a series of cutting passes. Clamp the stop firmly to ensure an accurate stop on every pass. (Some stops have feeble clamps and can be displaced by the slightest nudge of the saddle.)

FIGURE 1-41 Saddle stop with micrometer adjustment.

1-33 WHAT DO YOU INTEND TO DO WITH YOUR LATHE?

The three main "user categories" are these:

1. **Occasional use by the home-shop machinist.** Self-education and/or production of a few gadgets
2. **Miniature model making**
3. **Product development.** Experimental models for the engineering/physics lab, mold making, repair items, firearms, custom auto parts

No matter what the category, lathe size is the first consideration (Sections 1-1 and 1-2).

For the first category, *occasional use*, the other specs are a personal matter. It might be that a very basic used machine would be adequate. But equally, occasional use might develop into more serious lathe work, so your needs could expand into the "product development" category.

For the second category, *miniature model making*, there will likely be a requirement for screw cutting, but perhaps not for power-feeding. If you plan on cutting small screws, check the range of thread pitches available with the chosen machine. Only a few offer TPI numbers greater than 56 (#2 coarse thread). Similarly, the finest metric pitch is likely to be 0.4 mm (M2 coarse thread). So you need to ask, *How important is it to have those very fine threads?* Think for a moment about what's involved in machining small screws. The practical limit may not be so much the range of TPIs available with your lathe, but instead the difficulty of handling and cutting very thin bar stock, e.g., only about 0.1" diameter for a #4-40 thread.

ABOUT SMALL SCREWS . . .

A good question to ask is, *Why wouldn't I use a die instead of cutting them on the lathe*? Yes, you can (most machinists would do exactly that)—but there are times when it helps to ensure the squareness of the die by first cutting a partial-depth thread in the usual way with a single-point tool.

The third category, *product development*, likely calls for every imaginable lathe feature, including screw cutting and power feeding. Most machinists in this category will tell you that power-feeding both the saddle and cross-slide is essential, and will also expect a full range of screw threads, at least from 6 to 48 TPI (1-1/2" coarse to #4 fine), also metric pitches from 5 to 0.5 mm (M48 coarse to M4 fine).

CHAPTER 2

TURNING TOOL BASICS

CONTENTS AT A GLANCE

2-1 CUTTING FLUIDS AND SURFACE FINISH QUESTIONS

Although this might seem to have nothing to do with turning tools, cutting fluids are mentioned here because they can make a big difference to those starting out in lathe work. The basic fact is that you can machine aluminum, brass, and cast iron "dry," using no fluid at all. Some say that also applies to light cuts on mild steel. That's certainly true, but it may not always be good practice. Newcomers to lathe work should at least be aware of cutting-fluid options, and should be able to experiment with them to improve cutting speed, surface finish, and tool life.

The two main categories of cutting fluid are (1) "straight" oils, applied undiluted from the can (or little bottle), e.g., Tapmagic, and (2) water-miscible cutting oils, also known as "coolants," such as Blaser Swisslube. Coolants like Swisslube are used on practically all CNC machines to cut everything, including aluminum, and will probably be the first choice of fluid for any model shop lathe that comes with a fluid circulating pump and dispenser. See Section 4-10 for more.

The main function of a coolant is what you might guess—cooling the workpiece and cutting tool, thereby improving cutting action and saving wear on the tool. In the model shop you might use coolant when taking repeated cuts on steel, especially the harder alloys including stainless. However, you do not need a circulating system—use a *spray bottle* instead.

For routine work you might prefer straight cutting oil, applied with a flux brush as needed. Straight oil is certainly the choice for thread cutting with taps, button dies, or even single-point cutting tools. Try straight oil, too, when aiming for better surface finish, especially on steel.

Machining is always a voyage of discovery and surprises. Your favorite recipe of cutting tool, material, cutting speed, and lubrication that worked well yesterday seems not to work today. Something *must* have changed, but exactly what is anyone's guess. There is no good answer to this other than perseverance with the things we do have control over: tool sharpness and height, depth of cut, spindle speed, feed rate, and cutting oil—yes/no?

Finally, when machining is done, and you are still unhappy with the surface finish (especially on steel), consider using a fine file or a flexible abrasive pad such as 3M Scotch-Brite. The two grades I have found most useful are 7447 (very fine) and 7448 (ultra fine). If the workpiece is still on the lathe, protect the ways, chuck, etc., from abrasive dust.

2-2 BASIC TERMINOLOGY OF LATHE TOOLS

In one form or another, the right-hand (RH) knife tool is used in about 90% of all model shop turning applications. It is called RH because its cutting edge is at the right of the shoulder it creates in traversing the workpiece from right to left (Figure 2-1). Conversely, the LH (left-hand) tool is on the left of the workpiece shoulder. The angles of the various faces of these tools are stated with reference to the surfaces of the *tool shank*, usually a square cross section (if not, it helps to imagine it to be square).

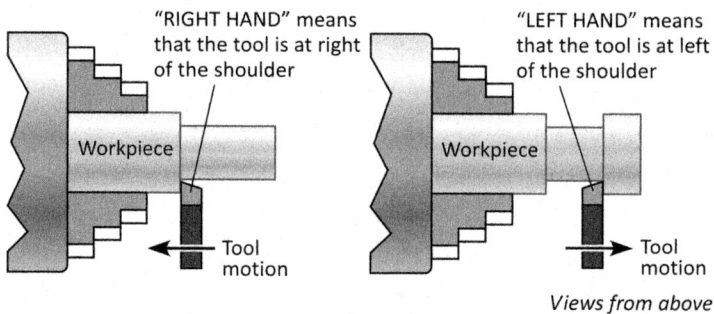

FIGURE 2-1   Right-hand and left-hand tools.

2-3 LATHE TOOL MATERIALS

It's assumed here that you have read the basics of metal removal in *How to Run a Lathe* or other sources. What has changed since the classics were published is the availability of hundreds of high-performance tools that make turning so much easier and faster. The two key factors are *high-speed steel* (HSS) and *tungsten carbide* (often shortened to "carbide"). Back in the day, the only metal available for cutting was high-carbon "tool steel," which is still used in vast quantities for wood chisels, knives, etc. (High-carbon tool steel is often preferred over HSS in applications where machining or shaping is called for prior to edge grinding, and where the tool is not subject to high temperatures from high cutting speeds.)

2-4 HSS SPECIFICS

Aside from carbide, high-speed steel is practically the only material used nowadays in the small shop for lathe tools, drills, milling cutters, reamers, etc. "High speed" means the steel tools can withstand the heat generated by fast, deep cuts better than yesterday's high-carbon lathe tools (which soften when overheated). HSS, which contains more exotic components such as chromium, molybdenum, tungsten, and vanadium, is a very distant relative of high-carbon tool steel. Unlike high-carbon steel, HSS is not machinable in the ordinary sense, so it has to be ground to the desired shape. HSS tool blanks are relatively inexpensive and readily available in square and round sections.

Another important difference is the fact that HSS can be ground just short of *dull red temperature* without affecting its performance in the slightest—a good thing if you are trying to grind, say, a single-point threading tool from a 3/8"-square blank of HSS. Compare that with sharpening the edge of a wood chisel, which was likely made from non-HSS tool steel: The slightest sign of discoloration (bluing) means that the tool is softened and won't keep an edge. HSS starts hard, ends hard.

Grind HSS as aggressively as you like to "almost dull red," *air-cooling* when necessary. No matter what you've read elsewhere, never cool HSS from high temperatures by water quenching. The sharp cutting-edge cools faster than the heavier sections of the tool, causing micro cracking at the edge, and possibly, premature failure in use. At lower temperatures quenching is unlikely to cause problems.

The most widely used formulation of HSS is M2, also known loosely as molybdenum, or "moly," steel (oddly, all high-speed steels contain molybdenum—don't look for logic here). The other HSS you occasionally come across is M42, which contains a high percentage of cobalt and is more expensive than M2. If you see "cobalt steel" in the catalogs, it is likely to be M42. Aside from its higher cost, M42 is more difficult to grind (more pressure, harder on the wheel), but it does hold up better at higher cutting speeds. This is not usually a factor in the model shop, so go with M2 if there's a choice.

2-5 SHAPING A LATHE TOOL

The basic requirement is a conventional two-wheel bench grinder, of the sort you probably have on hand. Grinders usually have one coarse-grit wheel and one fine. Don't use the coarse wheel unless you need to remove a lot of metal. It's difficult to be specific about grit size, because

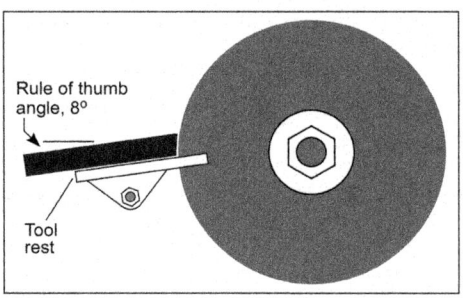

FIGURE 2-2 Grinding an end relief.

the label on many wheels doesn't say. (Who knows why?) Good choices for finish-grinding of a shaped tool are 60-grit or 80-grit aluminum oxide.

The grinder will need a flat tool rest along which you can slide the tool from side to side (Figure 2-2).

Purists will point out that the tool surface is concave ("hollow ground"), so there's really an infinite number of tangential grinding angles. However, we are not looking at exact science here. If the grinding wheel is the usual 6" or larger, and if the tool rest is about on the centerline, you have a close enough approximation.

Better than a bench grinder, according to some, is a belt sander, which in theory doesn't curve the ground surface. Unfortunately, it does, because the belt bunches up at the point of contact (worse, any curvature will be in the wrong direction—convex). Best of all is a dedicated tool grinder, but that's not necessary for first experiments. (If and when you do decide to invest the time, there are how-to instructions in Chapter 12.)

To illustrate the various angles of a lathe tool, we'll grind a right-hand knife tool from a square-section HSS blank, say 3/8" square. *Why such a big chunk of HSS?* The answer is . . . conventional thinking, nothing more (but it does make the illustrations here easier to read).

In practice, it takes much less effort to use 3/16" HSS blanks in a square holder (Section 2-11).

Square blanks often come with a beveled end to save you the effort of starting from scratch. If that's the case, it would be installed in the tool-holder as shown in Figure 2-3. (Surprisingly, even an off-the-shelf blank like the one shown in the figure will remove metal, not so much cutting it as pushing it aside.)

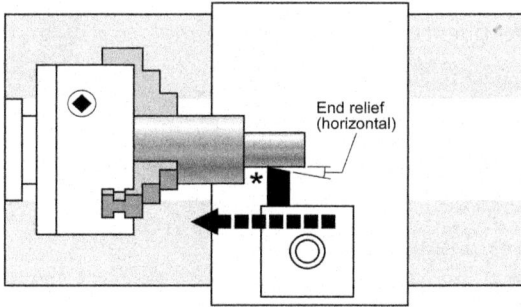

FIGURE 2-3   HSS cutting tool. The asterisk indicates the cutting edge.

2-6 END RELIEF

If your blank came with a square, unbeveled end, the first job is to make it as shown in Figure 2-3, which will stop it rubbing on the cut surface. Do this by grinding a *horizontal end relief*, say between 8° and 10°, as shown. While you are at it, angle the shank down to grind the *vertical end relief* at the same time, also in the range of 8° to 10°. The resulting ground surface forms a compound angle (Figure 2-4).

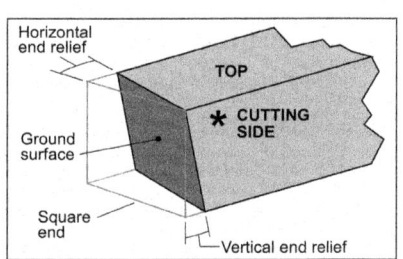

FIGURE 2-4   Ground blank with horizontal and vertical end reliefs. The surface of the "compound angle" shown here is idealized—in practice it would be inwardly curved by the grinding wheel, not flat. The asterisk indicates the cutting edge.

| 47 |

The compound angle thus ground may have improved the tool's cutting ability somewhat, because the "end relief" has taken care of most of the rubbing. But the cutting edge between the top and left sides is still a very blunt 90°.

2-7 SIDE RELIEF

To deal with the still-blunt cutting edge, we grind a *side relief* angle on the cutting side (Figure 2-5). With side relief the tool no longer rubs against the shoulder of the workpiece. However, it still lacks an efficient cutting edge. For that we grind a *side rake* on the top surface. (Yes, the terminology confuses everyone, but see Figure 2-6 for an explanation of the terms.) The result-

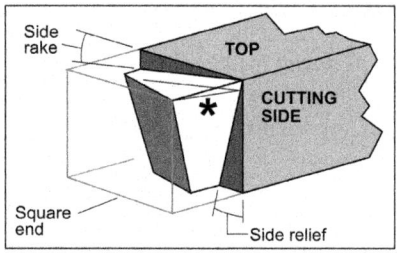

FIGURE 2-5   Finished tool with side rake and side relief. The asterisk indicates the cutting edge.

ing "angle of keenness" between the top and side surfaces will now be on the order of 70°—a very functional cutting edge for many materials.

Leaving aside the terminology issue, the cutting edge of a lathe tool is never defined by just one angle. Like every other knife, its keenness depends on the *included angle* between two surfaces.

Don't overdo the side relief angle. The larger the angle, the *less supported* (and therefore more fragile) the cutting edge.

The side relief grinding is done with the tool rest at the same 8° down angle as before (see Figure 2-2 shown earlier), but with the tool shank placed *across the rest*, pointing to the left. For the side rake, set the tool rest 8° up; then point the tool shank to the right.

2-8 DON'T WORRY ABOUT BACK RAKE

This is another complication to be aware of. Back rake (aka "top rake") tilts the top surface back by a few degrees, even as much as 20° for cutting aluminum. Now and again, it can be worth experimenting with back rake for faster cutting speeds, but for most purposes you can forget about top rake altogether. Note that Figure 2-5 shows no back rake.

2-9 TOOL ANGLE NOMENCLATURE

This can be confusing, to say the least. Figure 2-6 summarizes the generally accepted conventions, but don't be surprised by other opinions.

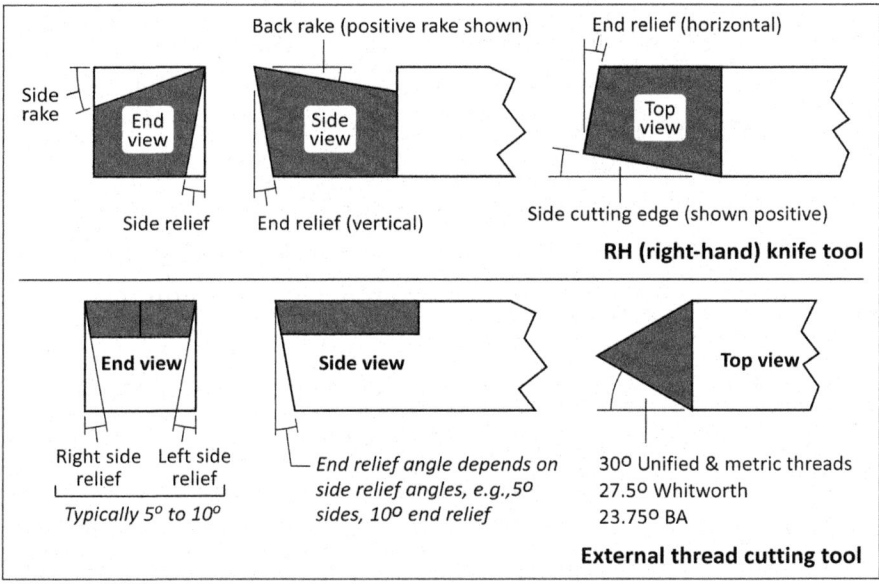

FIGURE 2-6 Cutting tool definitions.

Tool grinding angles depend on a number of factors, including personal preferences. There are no hard and fast rules, but here are a few general principles:

- **Side rake.** In general, the softer the material, the greater the rake angle should be. For low-carbon steel in the model shop, you might use a side rake angle of 10°, but harder alloy steels cut better with a rake of 5° or even lower. Aluminum cuts best with a rake angle of 20° or more, even as high 30°.

- **Side relief.** For general work with HSS tools, a side relief angle of 10° is recommended. Harder, more brittle carbide tools survive better with a smaller side relief angle, providing more support under the cutting edge. Similar numbers apply to the end relief angles.
- **Side cutting edge angle.** This affects the thickness of chip removed by the tool; the larger the angle, the thinner the chip. (With a side cutting edge angle of zero degrees, the chip thickness is the same as the feed rate, expressed as some number of thousandths per revolution.) The side cutting edge angle of shop-ground tools is usually zero, so any changes necessary are achieved by rotating the QCTP. Rotate the tool post clockwise for a positive side cutting edge angle. Often, we want to rotate the other way, for a slightly negative side angle, so we can cut and face with the same setting (see Figure 2-15).

Beware of large side cutting edge angles, in either direction. A negative angle, for the two-in-one cutting/facing function, pulls the workpiece toward you. Conversely, a positive angle pushes the workpiece away. These forces have the effect of distorting the workpiece, especially if thin, and also of taking up backlash in the cross-slide lead screw. These are the leading causes of unexpected, unwanted taper turning. (*Hint:* You can take care of one of these issues—backlash—by locking the cross-slide before each cutting pass.)

2-10 TOOL HEIGHT

Tools cut better if their cutting edges are at the same height as the lathe's longitudinal axis, in other words, the center height. This applies particularly to facing tools. With today's QCTPs, this is much less of a problem than it used to be (Section 1-28).

2-11 THINK SMALL

How *long* does the cutting edge of a lathe tool have to be? This depends, among other things, on the anticipated depth of cut. In the model shop, we try not to take anything as big as 1/4" off the radius, and then only if the workpiece is aluminum or plastic. More likely, we will be thinking of 1/10" or less.

It follows that all we need for a cutting edge is one that spans just a little more than our maximum cutting depth. So for shop-made HSS tools, think *small*—bits rather than complete tools with their own shanks. You will save a lot of time, effort, and wear on the grinding wheel by starting with a 3/16"-square or 3/16"-diameter M2 blank. These are available from all tool suppliers for a very few dollars apiece. Cut the blanks to length using a Dremel cut-off wheel. Two examples are shown in Figures 2-7 and 2-8.

FIGURE 2-7 Small bits are easier to grind. This is a 3/16"-diameter HSS rod sharpened as a right-hand knife tool, *enlarged view on the right*. It is held in a shop-made 1/2"-square block installed in a QCTP toolholder. In the foreground is a second 3/16" rod with a 60° point for use in thread cutting.

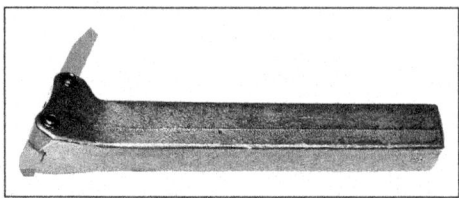

FIGURE 2-8   Commercial holder (Wimberley) for 3/16"-square HSS. This has a built-in rake in two axes. Very little grinding is needed to achieve an efficient cutting edge, capable of facing, reducing diameter, and turning to a shoulder without adjustment.

2-12 HONE THOSE EDGES

This is a key part of grinding your own tools. Provided you use 60 grit or finer, the edge left by the grinding wheel may be usable as is. For a keener, longer-lasting edge, use an oil stone or a diamond lap (Figure 2-9), but hold it *flat*; don't "round over" the edge. This will make the tool worthless, and it's only too easy to do.

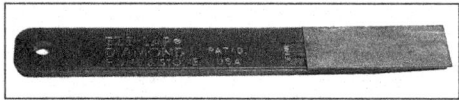

FIGURE 2-9   Diamond lap. Use a diamond lap or oil stone for the finishing touch and also for resharpening a used tool.

One thing that *does need rounding* is the tip of the tool. As it comes from the grinding wheel, the tool tip is needle-sharp and won't last a moment on anything other than plastic. Aim for a tip radius on the order of 10 to 15 mils, a shade less than 1/64". This will deliver a much smoother surface finish.

2-13 TUNING UP THE GRINDING WHEEL

This is called "wheel dressing," the process of removing embedded metals (glazing), ridges, valleys, and rounded-over edges from the cutting face of the wheel. Imperfections like these make it difficult to grind a flat surface. (Glazing looks just like it sounds—a shiny surface that doesn't do anything but heat up the tool.)

The usual form of wheel dresser is a heavy cast-iron handle with a stack of three or four star-shaped hardened rotors at one end (Figure 2-10). Run the grinder up to speed; then press the rotors against the wheel, traversing from side to side as necessary to achieve a perfectly clean, square edge.

Wheel dressing is an ugly process, best done outdoors to keep the clouds of abrasive dust away from anything you value—especially machine tools.

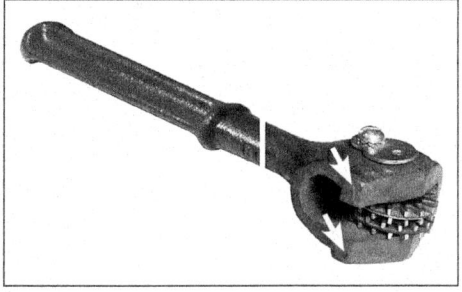

FIGURE 2-10   Grinding wheel dresser. Set the feet (to which the arrows are pointing) on the grinder tool rest, apply forward pressure, and then traverse the dresser across the wheel.

2-14 TUNGSTEN CARBIDE

Tungsten carbide is often assumed to be a metal. It is actually a *compound* of a metal: tungsten and carbon. The result is a powder that is combined with another metal, cobalt, to form solid shapes for use as cutting tools. Carbide tools are extremely hard—about 75 on the Rockwell C scale, compared with 60 to 65 for high-speed steel.

Carbide would be used for every conceivable tool on the planet if it were not for one downside: poor impact resistance, causing the tool to fracture if it nudges the workpiece too briskly. Additionally, it is too hard to be sharpened on a regular grinding wheel—use a diamond wheel instead.

The good news for the small metal shop is that carbide-tipped tools are readily available, at more or less sensible prices.

So why bother with tungsten carbide when HSS works just fine?

The answer: *Carbide holds up better when machining hard materials and doesn't need resharpening so often.* It allows (actually works better with) higher cutting speeds than does HSS.

However, though the cutting edge of a carbide tool does indeed stay keener for longer, it's also fragile. It can fracture in an instant, especially with "pointy" tools such as thread cutters—so think *gentle contact.*

Another issue is surface finish: A nicely ground, round-tipped HSS tool can often give better results, with less cutting pressure, than a similar-looking carbide.

The aluminum oxide wheel used for HSS doesn't work on carbide. Use a green silicon carbide wheel or (better) a diamond wheel instead. (The downside of silicon carbide is that it breaks down quickly, delivers a marginal cutting edge, and creates a dangerous amount of dust.) A diamond wheel works beautifully, but its cost puts it out of range for the small shop (unless you have made a dedicated tool grinder, a subject addressed in Chapter 12). For anything other than *occasional resharpening* with a stone or diamond lap, the practical answer is usually to buy new.

One point to bear in mind: HSS usually delivers better surface finish than carbide. Also, in general, choose HSS over carbide for aluminum work.

2-15 GRADES OF CARBIDE

At the industrial level, there is a high degree of fine-tuning carbide grades to specific applications. This doesn't apply in the model shop, where the choices usually come down to C2 and C6 (occasionally C5)—and that's if the grade is mentioned at all. Fortunately, carbide tools, especially imports, are inexpensive enough to take a chance that the supplier has it right.

After many years of using carbide tools, I still can't reliably differentiate between one grade and another. I have always purchased the "steel-compatible" grade (because that's mostly what I machine), but I have used the exact same tool on every other material, including aluminum, brass, Delrin, and PVC.

For nonferrous metals, including aluminum, brass, and bronze, also cast iron, go with C2 (said to be less likely to fracture than C6). For steel, choose C6 (or C5). This is the harder of the two grades (holds its edge better).

2-16 TUNGSTEN CARBIDE TOOL FORMATS

There are two main classes:

1. Factory-made carbide-tipped tools
2. Indexable carbide tools

Each of these (respectively) is the focus of the next two sections.

2-17 CARBIDE-TIPPED TOOLS

These are an inexpensive start-out option. Many users never use anything else—other than occasions when they need to grind their own special shapes. These tools are manufactured by brazing a chunk of solid carbide onto a square steel shank, which can be anything from 1/4" to 1". Figure 2-11 shows the more popular styles, and Figure 2-12 shows real-life variants.

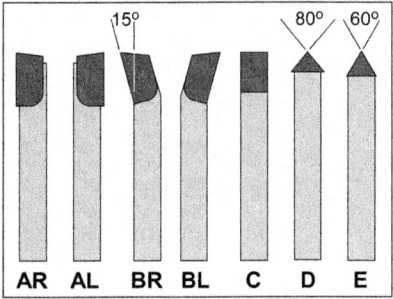

FIGURE 2-11 Carbide-tipped tool formats.

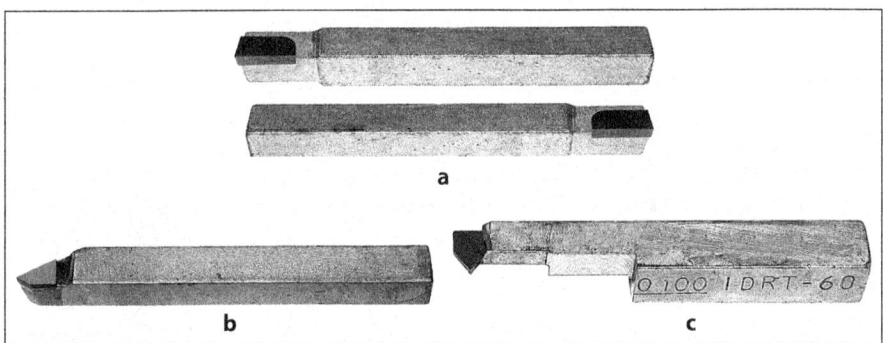

FIGURE 2-12 Carbide-tipped tool examples (a) Knife tools Style A RH upper, LH lower. (b) Thread cutting tool Style E. In this example of Style E (60° included angle), the tip is offset to allow thread cutting nearer to a shoulder. (c) Internal thread cutting tool.

Three things to bear in mind:

- *First*, carbide-tipped tools are not optimized for specific materials, so they don't offer a choice of rake and relief angles (side rake, side relief, and the two end relief angles are usually between 7° and 8°, a good general compromise).

- *Second,* you rarely need Styles B, C, or D, so don't be tempted to buy a set. You can achieve the effect of Style B simply by rotating the tool turret or quick change tool post a few degrees (rarely as much as 15°). (I have never found a use for Styles C and D, but I often use a 60° tool like E for thread cutting if there happens to be a sharp one in the box.)
- *Finally,* every style of carbide-tipped tool you buy needs its own QCTP toolholder—otherwise, you will spend a lot of time shimming for center height.

Sharpen carbide-tipped tools with a diamond lap, taking care not to round over the cutting edge.

2-18 INDEXABLE CARBIDE TOOLS

These are my first choice for routine turning (Figure 2-13). I have just two of them, a left-hand and right-hand pair of 1/2"-square tool shanks, plus the smaller one shown in Figure 2-16 (later in the chapter). I bought them years ago from McMaster Carr. (In the company's current catalog, they are referred to as "economy indexable turning tools.") When I first bought them, these inserts sold for about $3 each. At 2022 prices they go for more than $7, so much for economy. You can still buy inserts costing less than $4 each from other suppliers, but be sure they fit your tool shanks—they may look alike, but there are many variants out there.

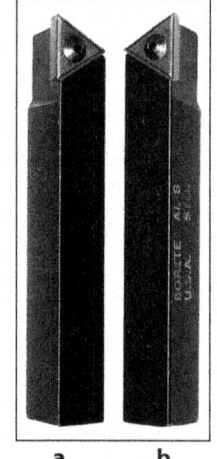

The inserts are triangles of solid carbide, mostly with a chip-breaker channel on each edge (Figure 2-14). Like carbide-tipped tools, they are available in C2 and C5/C6 grades (Section 2-15). Inserts are secured to the tool shank with a special tapered screw.

Indexable inserts can save a lot of time, effort, and frustration. When the cutting edge dulls, simply loosen the screw and then rotate the insert to expose a fresh

a   b

FIGURE 2-13
Indexable carbide tools. (a) RH knife tool. (b) LH knife tool.

edge. When all three are exhausted, discard the insert, or try honing the edges if you don't have a fresh insert on hand.

The tip radius of an insert is important (Figure 2-14). On the one hand, a 1/32"-radius tip fractures less readily than a 1/64" tip—good to know. On the other hand, it leaves a larger fillet at the transition to a shoulder (Figure 2-15). The fillet can usually be removed by pushing the cross-slide in a shade over 10 thousandths.

Indexable carbide knife tools, as shown in Figure 2-13, are only a very small part of the story. As you can see from a glance at any machine tool catalog, there are literally hundreds—if not thousands—of indexable carbides on the market. The majority of them are specialized inserts for industrial use, often at incredibly high prices.

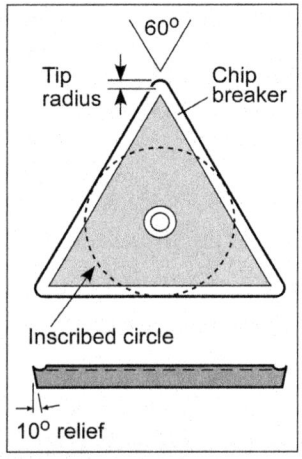

FIGURE 2-14 Indexable carbide insert. When purchasing, specify the inscribed circle (usually 1/4" or 3/8") needed for your tool shanks. Also specify the desired tip radius, usual choices 1/64" and 1/32".

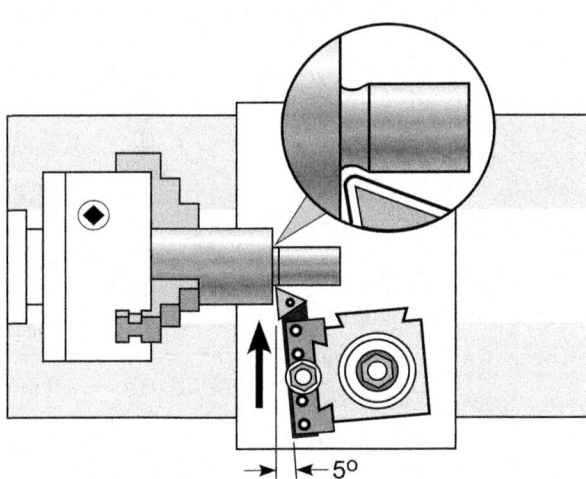

FIGURE 2-15 Removing a shoulder fillet caused by tip rounding.

2-19 SCREW CUTTING TOOLS

Screw cutting on the lathe comes with a host of special conditions and caveats. These are covered in Chapter 5. Here we look only at the cutting tool itself.

Unified and metric threads are cut with a 60° tool, possibly carbide-tipped, as Style E in Figure 2-10 (shown earlier). In a pinch, you could use a carbide insert (Figure 2-16), but its tip radius is usually too large unless you are cutting a very coarse thread.

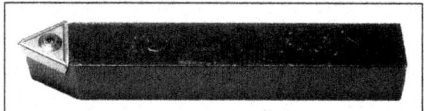

FIGURE 2-16 Carbide insert on a special tool shank.

FIGURE 2-17 HSS screw cutting bit ground from 3/16" rod.

Better than a standard triangular insert would be a small HSS bit, as shown in Figure 2-17, ideally ground so that it can be *flipped upside down* without needing height adjustment. "Flippability" is very helpful when switching from right-hand to left-hand thread cutting (and other configurations), a hint of which is given in Figure 2-18.

I have had good results over several years with special screw cutting inserts such as the one shown in Figure 2-19. These too are flippable because the centerline of the insert is on the centerline of the tool shank. They are double-ended, so there is a fresh cutting tip available if the first

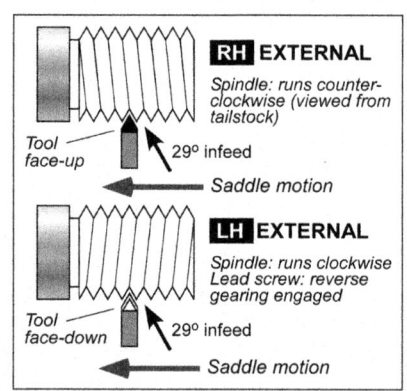

FIGURE 2-18 Invert the tool to cut left-hand threads. This allows you to use the same 29° compound setting and infeed direction as for right-hand threads. Tool (saddle) travel is from right to left in both cases. See Chapter 5 for more options.

one breaks off—a rather frequent occurrence. Another plus for this tool is its ability to cut very close in against a right-hand shoulder.

2-20 INTERNAL CUTTING TOOLS

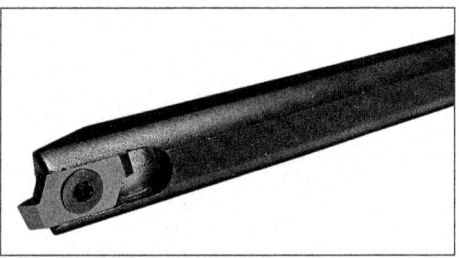

FIGURE 2-19 Right-hand thread cutting insert (courtesy of Mesa Tool, author photo).

These tools are a source of so much confusion that they need to be considered separately. Regular "external cutting," which we do 99% of the time, is a breeze compared with internal cutting. For one thing, there is virtually no limit on the size of an external tool, whereas a boring tool by definition has to fit inside the bore. This means a trade-off between the tool shank diameter and the rigidity we need at the cutting edge. "Springiness" may cause the tool to cut a different diameter on the in-stroke versus the out-stroke—beware of this when checking bore size.

Most internal cutting tools are small HSS or carbide bits (inserts) held in "boring bars" usually 3/8" diameter or more.

If the tool shank (i.e., the boring bar) is 3/8" diameter all the way to the insert, as shown in Figure 2-20, the smallest internal diameter it can cut will be about 1/2", and that only with the tiniest amount of cutter bit pro-truding—a lot less than that shown in the diagram. This is one reason for the popularity of brazed carbide-tipped boring tools, as shown in Figure 2-21. Because their stems are relieved for separation from the bore, these tools allow bore sizes as small as 5/16", even with a shank diameter of 3/8".

However, no matter what the design of the tool, there is always the question of how to hold it rigidly. Because I have only a handful of boring tools with square shanks—which are easier to grip in the QCTP—I have had to come up with custom shop-made adapters, examples of which are shown in Figures 2-22 and 2-23.

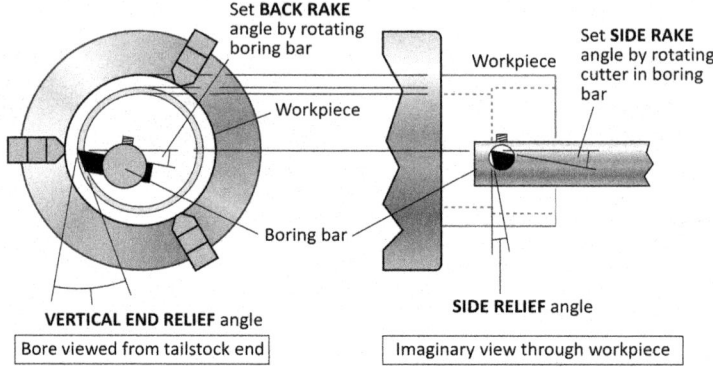

FIGURE 2-20   Boring tool angle nomenclature.

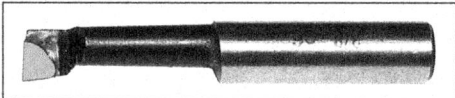

FIGURE 2-21   Carbide-tipped boring tool in a
3/8" shank.

FIGURE 2-22   QCTP toolholder
with 3/8" ID adapter for HSS
boring bit.

FIGURE 2-23   HSS insert in long tool shank. Shop-
made heavyweight toolholder with 3/8″ bore.

2-21 ARE THERE CARBIDE INSERTS FOR INTERNAL CUTTING?

Yes there are, but they aren't often found in the small shop, the main reason
being over-the-top prices. However, if you need to bore an ID *as small as
1/4″*, you might want to take another look. There are a few *miniature inserts*

with thin stems that can do such jobs. They are indeed pricey, but they do take care of an otherwise impossible task.

2-22 IS THE HOLE THE RIGHT SIZE?

For noncritical holes, especially small ones, we simply drill the hole with the appropriate-size drill bit and don't give it another thought. Sometimes we might use a reamer for greater precision. For larger holes that are cut with a single-point boring tool, we need special gauges (see Section 4-35).

2-23 GRINDING INTERNAL TOOLS

If they're not brazed to the tool shank, cutter bits for boring are typically round cross sections such as broken 3/16" drills or other HSS remnants. The boring bars themselves are also usually round, so you can have a virtually unlimited range of angles simply by rotating the cutter bit in the bar (side rake/side relief) or by rotating the boring bar (back rake/end relief—see Figure 2-20).

This means that *side rake* and *side relief* have to be considered as a pair (if you change one by rotating the bit, you change the other), so you need to think of the difference between them—the "angle of keenness"—instead of two specific angular values. A good number to start with is a difference of 70°.

Vertical *end relief* and *back rake* are also a pair. The bottom line is that the end relief face must be ground so that it doesn't rub on the workpiece, no matter what the back rake and no matter what the bore diameter. Grinding tool bits for regular "external cutting" is straightforward because the effect of each grinding operation is easy to visualize. Not so with boring bits. For one thing, the bits are tiny; for another, it is not easy to see which ground face does what.

You can grind freehand, which I do most times with carbide-tipped tools. For too-tiny-to-hold HSS bits, I use a square section holder such as the one in Figure 2-24. This is large enough for easy handling, meaning it can slide nicely back and forth along the grinder rest. With a holder like this, the cutter bit becomes just another lathe tool, just as easy to grind (it is, in effect, a left-hand knife tool).

2-24 CUT-OFF TOOLS

Cut-off tools (aka parting-off tools) are in a separate category of their own, not because they cut into the work-piece at right angles, but because they

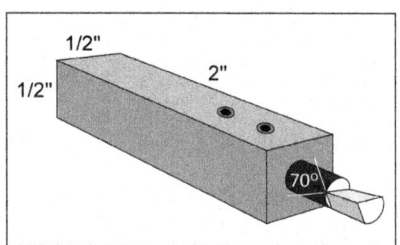

FIGURE 2-24 Holder for grinding 3/16"-round HSS boring bits.

cause more problems than all the other tools put together. Cut-off tools are used for grooving and for separating a finished-turned part from stock held in the chuck.

The "standard" QCTP cut-off toolholder is shown in Figure 2-25. The 100 Series toolholder shown here is designed for blades 1/2" tall. HSS cut-off blades are available in various thicknesses from 0.04" to 1/8". Not all cut-off blades have rectangular cross-sections. Some are T-shaped, with a cutting edge wider than the shank (see the inset in Figure 2-25). These blades need to be held vertical to avoid rubbing in the groove. Use a shim, in this case 0.02", between the shank and toolholder.

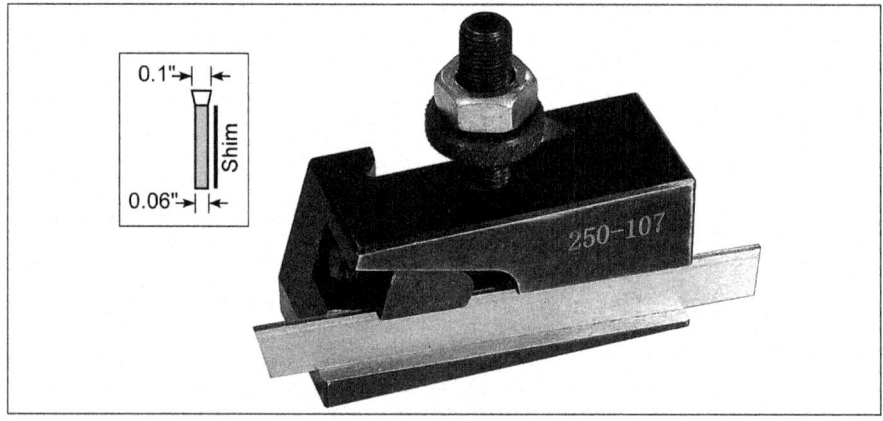

FIGURE 2-25 Typical QCTP cut-off holder with HSS blade. The cut-off toolholder is angled at 4°, so there is usually no need for additional grinding on the top surface of the blade. Shown inset: a representative T-shaped cut-off blade.

One factor to bear in mind when choosing a blade thickness is material wastage—a 1/8"-thick slice is a lot to lose. So why not go with 1/16"? One answer is that a 1/16" blade is so flexible that convex or concave cut-offs are more than likely. The same applies to 1/8", but to a lesser extent.

For suggestions on dealing with cut-off problems, see Chapter 4. Meanwhile, this chapter would not be complete without mention of a highly effective cut-off blade only recently available at sensible prices— namely, the carbide-insert cut-off, Figure 2-26. These are mostly from China, surprise, but in this instance. there seems to be interchangeability between manufacturers; don't bank on this though—buy the holder, the blade, and the inserts at the same time from the one supplier. All three items were available in 2021 from eBay suppliers for a total of about $50.

As supplied, the blade holder does not fit a standard 100 Series toolholder, but that can be taken care of in a few minutes of mill work. Compared with every other scheme I've tried to deal with the cut-off problem, this one is a breeze. It cuts square, without lubrication, at high speeds; and because the insert is wider than the supporting blade, it doesn't jam in the cut. Two more bonus features: (1) Because the blade is not angled, it can be extended without needing height adjustment. And (2) a replace-

FIGURE 2-26 Carbide-tipped cut-off tool and inserts. Dimensions of this model: blade height 26 mm (1.02"), thickness 1.6 mm (0.063"), insert thickness 2 mm (0.08").

ment insert can be installed without removing the holder from the tool post.

2-25 GROOVING TOOLS

Grooving tools are similar to cut-off tools in that they cut into the workpiece at right angles. But that's it for similarity. Often used to cut grooves for retaining rings such as circlips and E-rings, grooving tools are much thinner than any cut-off tool. For example, a retaining ring for a 1/4"-diameter shaft needs a groove only 0.029" wide. There may be inserts or blades of that thickness, but I have never come across them.

What I use is shown in Figure 2-27. This is a blade measuring 1.45 mm x 5 mm (about 0.06" x 0.2") in a 3/8"-square shank, which is clamped in a standard QCTP toolholder. Grizzly at one time carried this item, calling it a "parting tool with blade." The blade tip in the photo was thinned to a little less than 0.029". It is rigid enough to be run in repeatedly to widen the slot as necessary.

The good news with grooving, as compared with cutting off, is that the depth called for is only a few thousandths, 0.01" for a 1/4" shaft, 0.015" for 1/2".

FIGURE 2-27   Mini cut-off tool used for grooving.

2-26 DRILL BITS

There's really no way around it. Every machine shop needs a sizable complement of drill bits, number sizes from #1 to #60 (0.228" to 0.04"), letter sizes from A to Z (0.234" to 0.413"), and, of course, fractional sizes from 1/16" to 1/2".

Drill bits are one of the few items of shop equipment you might want to consider buying in sets. No matter how you buy them, be sure to specify only high-speed steel. Most of them are high-speed today, even the low-priced imports, but you need to be sure. (If you don't see "HSS" on the tool shank, it is likely to be ordinary high-carbon tool steel, which doesn't hold an edge reliably, especially if overheated.)

No-name import HSS drills (along with most other HSS cutters) are usually good for model shop use, but may not hold up in high-volume production.

Drills are often offered with various coatings for longer life, faster cutting, etc. At the model shop level, it's difficult to see the difference, and it probably doesn't matter enough for a second thought. You can see in Figure 2-28 that where most of the drills in a set are plain, the odd one or two—replacements—are coated with black oxide and vice versa. Black oxide does provide a degree of self-lubrication and comes at no significant cost (unlike more exotic coatings).

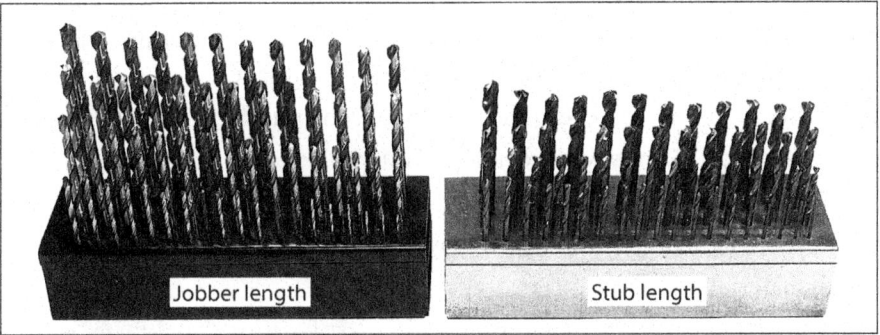

Jobber length
Stub length

FIGURE 2-28 Drills compared. Both of these are sets of "number size" (drill gauge) drills. Stub length is the better choice for most machining operations, but having both on hand is useful when a longer reach is called for.

2-27 DRILL LENGTH

The standard-length drill bit stocked by hardware stores is known as "jobber length" (who knows why?). More useful for the model shop, especially in small diameters, are stub-length drills (aka "screw-machine length"). These shorter drills *flex much less* and can sometimes be used without the need for preliminary center drilling.

How do jobber length and stub length compare? Jobber-length size #1 (0.228") is 4" long, and stub-length size #1 is 2-1/2". My jobber-length drills are used only for deep holes, so they gather dust most of the time.

2-28 DRILL GEOMETRY

There are two main classes of drills used in the machine shop, differentiated by *point angle*, usually 118° or 135°. General-purpose jobber-length drill bits from the hardware store almost always have a 118° point angle (Figure 2-29).

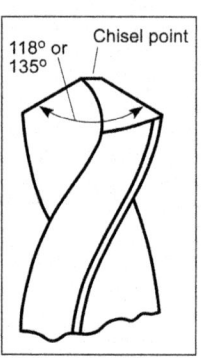

FIGURE 2-29
Drill-point angles.

Both 118° and 135° drills can be used on practically all materials, but the general rule is that the sharper the angle, the better it is for softer materials. Many machinists prefer the 135° point angle, which is said to deliver a more truly round hole that's closer to nominal size. That may be true, but in regard to hole diameter I have never been able to tell the difference between brand-new 118° and 135° bits when drilling mild steel.

However, if you have 118° and 135° bits of the same size and quality on hand, you will definitely notice a difference between the two when drilling a tough material such as stainless steel. The 135° drill cuts to diameter faster and is less apt to walk off center.

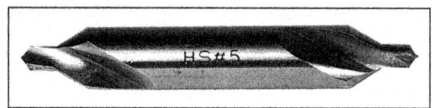

FIGURE 2-30 Drill-point countersink with two 60° tips.

That said, my advice is to *center-drill* beforehand every time, no matter what the material or the point angle (Figure 2-30). You will need two or three sizes of *drill-point countersinks* with body diameters of, say, 1/8", 1/4", and maybe 5/16".

Double-ended bits are classified by "trade size," but that is not a complete description. Each trade size is available in several lengths, and the smallest sizes (0, 00, etc.) come with various tip diameters. The most common sizes for the small shop are listed in Table 2-1.

TABLE 2-1 Most common sizes of countersink bits for the small shop

Trade Size	Body Diameter	Drill Tip Diameter
0	1/8"	1/32" (0.031)
1	1/8"	3/64" (0.047)
2	3/16"	5/64" (0.078)
3	1/4"	7/64" (0.109)
4	5/16"	1/8" (0.125)

The most frequent use of the 60° countersink is "spotting" a pilot hole for further drilling. It is also used to prepare a workpiece for turning between centers (discussed in Section 1-15). See Section 4-37 for other countersinks.

2-29 SPLIT-POINT GEOMETRY

This is one more complication that applies mostly to 135° point angle bits. Until recently, 135° bits looked just like the general-purpose 118° variety, but with a less pointy tip. Today, all 135° drills, even small number sizes (e.g., #60, 0.04"), are available with split points (but not all suppliers stock them—you need to ask). Split-point drills (Figure 2-31) cut with about

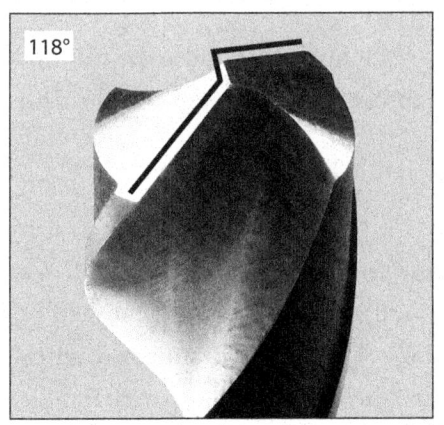
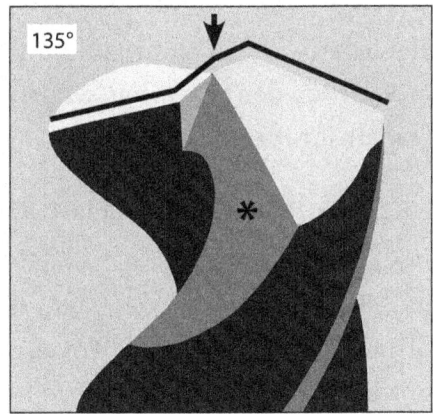

FIGURE 2-31 Drill-point geometry comparison. The 118° drill has a standard chisel point. On the 135° drill, the chisel point has been split by two secondary grinding operations: (1) the asterisked surface shown here, and (2) its counterpart behind the drill. Cutting edges are outlined in black. There are three cutting edges on the 118° drill and four on the 135° drill. The "split point," indicated by the arrow, forms a better-defined, smaller center pivot.

50% less thrust than standard drills and also stay on point more reliably. So *what's the downside?* Answer: *Sharpening.*

Split points are out of range even for the most skillful toolmaker to grind by hand (yes, people still do that sometimes on regular drills). Doing it right takes a dedicated machine with a fine-grit diamond wheel. I don't have such a thing, so I send drills out to a sharpening service. Either that, or I buy new if it would cost less than regrinding.

2-30 DRILLING SMALL HOLES

For less flexing under pressure, use stub-length drills if you have them. Back the drill out frequently for *lubing.* Run the spindle faster than you might think reasonable. *Example:* According to *Machinery's Handbook,* mild steels such as 1018 should be drilled with a cutting speed of about 100 feet per minute; that's 2000 rpm for a 3/16" drill, 1500 rpm for 1/4". For anything smaller than 3/16", the "right" speed is simply not available on the average lathe—so run it as fast as possible. Consider a *tailstock drill* for better pressure control and easier withdrawal for lubing (Section 3-10).

2-31 CONTROLLING DRILL DEPTH

The best answer is to install a digital readout on the tailstock (Section 1-29). If you don't have one, set the depth using adhesive tape on the drill, or mark it with a fiber-tip pen.

2-32 DEBURRING A DRILLED HOLE

Use a standard 82° or 90° countersink bit with a 1/4" shank held in a so-called adjustable pin vise. This is actually a keyless chuck mounted on a handle (see Section 3-12). For larger holes I use a 5/8" diameter counter-sink bit, which is beefy enough to be hand-held without a separate handle.

2-33 DRILLING LARGE HOLES

Don't try to drill them in one go. Bad things happen when a lot of pressure is applied to a drill bit. Anything above 1/4" goes better with a full-depth pilot hole, about 1/3 of the final diameter. Use a *much slower speed* when enlarging a hole. *Example:* Drill a 1/4" hole at the regular speed; reduce to 1/2 speed or less when enlarging to 3/8" or 1/2".

How do you stop large drills from rotating in the tailstock chuck? Often, the tailstock taper slips, and the chuck itself rotates. *Bad!* The drill shank, the chuck taper, or the tailstock's internal taper, or all three, can be mangled.

2-34 WORKING WITH OVERSIZE DRILLS

Reduced-shank drills, aka Silver & Deming drills, are available in sizes up to 1-1/2" diameter, with 1/2" shanks (Figure 2-32). Round shanks can slip in the tailstock chuck, so flatted shanks (three flats at 120°) can sometimes be a helpful option.

FIGURE 2-32   1" Silver & Deming drill with 1/2" shank.

A word of caution: Large drills like these put a heavy load on the lathe and the workpiece, which is why it is important to follow these three rules:

- **Rule #1** is to run the spindle at a *very low speed*, at least to start with.
- **Rule #2** is to grip the drill using padded locking pliers, resting the pliers on any solid surface that will allow the pliers to slide freely as the drill goes deeper into the work (Figure 2-33). This stops slipping in the chuck (which chews up the drill shank) and—equally important—prevents the chuck's taper shank from rotating.
- **Rule #3**, if necessary, is to stop the drill from jumping around and chattering by clamping a thick scrap of plastic or wood in a QCTP toolholder; then advance the cross-slide to press the material *gently* against the side of the drill.

FIGURE 2-33 Stop large drills from rotating. This 3/4" Silver & Deming drill is held by locking pliers, which are free to slide along the tool-post base. A leather pad prevents marring of the drill shank.

2-35 DRILLING DEEP HOLES

The deeper the hole, the greater the accumulation of chips in the flutes. This calls for frequent withdrawal of the drill for cleaning and lubing—increasingly time-consuming because it calls for more and more turns of the handwheel, both in and out. A fix for this is to unlock the tailstock, then slide the tailstock right to clear the workpiece (but make sure the chuck taper doesn't unseat). Keeping the drill in the clear, retract the tailstock quill; then slide the tailstock gently forward to re-engage the drill. Repeat the slide forward/slide back routine until the hole is drilled to depth. Another possibility is to use a tailstock drill (Section 3-10). When drilling deep holes in plastics that tend to squeeze down on the drill, use progressively smaller drills as you go deeper. When you are to depth with a small drill, finish the hole with the right-size drill.

2-36 DRILLING A FLAT-BOTTOMED HOLE

Predrill close to final size in the ordinary way; then finish with an end mill in the tailstock chuck. It is not necessary to use a center-cutting milling cutter, provided the starting hole is only a small amount undersize.

2-37 SCREW CUTTING BASICS

Screw cutting, aka "thread cutting," is a multilayered topic, so much so that there is a separate chapter devoted to it, Chapter 5.

On one level, you might say that every *commercially made* 1/4-20 screw fits every nut of the same spec, and that's largely true. Most of the time, we get by just fine, because all we need is for screw A to fit nut B, and we don't mind that nut C is a bit tight. Shop-made threads can be a different matter, especially when we are working to specific dimensions. It's at those times that we need to pay closer attention.

In the shop, threads are cut in three ways:

1. Single-point cutting tools for both internal and external threads (Section 2-19)
2. Taps for internal threads
3. Dies for external threads

2-38 TAPS

Aside from physical differences in flutes and chamfering (Figure 2-34), taps come in different *pitch diameters* designated H1, H2, and so on (often prefixed with "G," meaning "ground threads"). The higher the number, the larger the diameter (for metric threads, "D" is used instead of "H"). What this means is that an H1-tapped hole might be an impossibly tight fit for a screw you just made using a single-point cutting tool or a button die. (This may be no more than a temporary setback, provided the screw is still on the lathe—slim it down a thousandth or so by making another pass with the cutting tool. If you used a die instead, tighten it; then run it down the workpiece again—more on that in Section 2-40.)

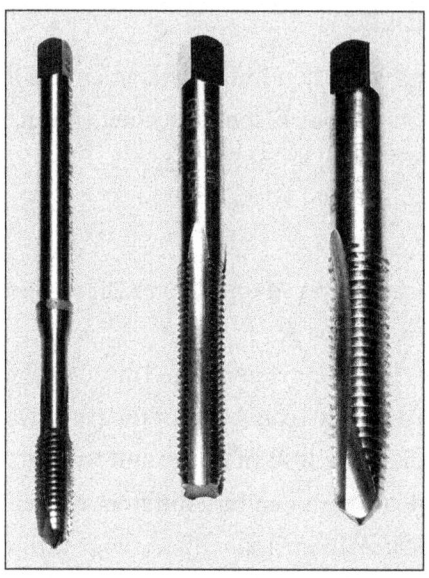

FIGURE 2-34 Taps. Aside from the obvious—nominal size and number of threads per inch—taps come with other visible differences in the number of flutes and lead-in chamfering. (There are also nonobvious variations in pitch diameter.) The examples here are (*left*) 3-flute "plug chamfer" (4 chamfered threads), (*center*) 4-flute "bottoming chamfer" (2 chamfered threads); and (*right*) 3-flute "taper chamfer" (7 chamfered threads, with a scalloped point). The scallop is intended to eject chips ahead of the cutting action—very effectively, but only on through holes.

Some suppliers do not say or even know what H number they are offering, so you may need to ask. A middle-of-the-road choice is GH3. HSS is a must for taps. Don't buy sets unless you come across a great deal. You will rarely, if ever, need a fully tapered tap—in fact, you can usually get by with only a bottoming tap if the hole is first countersunk a little with a center drill.

There is a lot more on cutting and measuring threads in Chapter 5.

2-39 HOLE SIZES FOR TAPPING

There is nothing sacred about the hole sizes recommended in the usual thread tables. They are designed to give you a 75% depth of thread, but 75% isn't a magic number. According to industry sources such as Greenfield Tap and Die, there is little gain in strength over a 60% depth. This means you can drill one or two number sizes larger than shown in the tables. *Result: Less chance of tap breakage*, less effort, and less slippage of the workpiece in the chuck.

2-40 TAPPING TECHNIQUES

There are, or there used to be, tail-stock tap holders such as shown in Figure 2-35. Basically, this is a conventional tap holder with the addition of a parallel stub that slides in a sleeve held in the tailstock chuck. This allows the tap to be fed several turns into the work without any drag from the tailstock. For screw sizes smaller than #10/M5, I always *hand-turn* the spindle when tapping.

FIGURE 2-35 Sliding tap holder for the tailstock.

No matter what, don't forget the cutting oil when tapping steel!

You can get by without a sliding tap holder (I do routinely), but be aware that small taps are very fragile and can break in the workpiece. Furthermore, whatever you may have read elsewhere, broken taps can rarely be removed in the small shop. In other words, start over.

Other ways to hold the tap: (1) For smaller taps, I use a tailstock drill press (Section 3-10). (2) For most M4 or 8-32 taps and larger, I typically use the regular tailstock chuck, easing the (unlocked) tailstock along the bed by hand to relieve drag on the tap. (3) For larger taps with a center hole at the end of the shank (M8 or 5/16"+), I use a tailstock center for alignment while holding the tap with an open-style hand wrench (Figure 2-36).

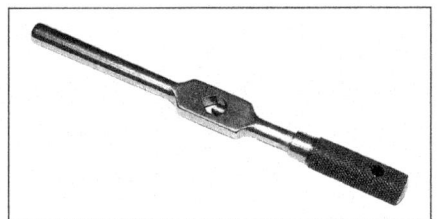

FIGURE 2-36 Open-style Starrett tap wrench.

No matter which way you go, the procedure is similar—hold the tap firmly, and then advance the tap gently into the hole by sliding the tailstock or by moving the drill press lever forward. Maintain a light pressure on the tap—helping the tap move forward—while rotating the spindle by hand. Every turn or so, back the spindle to clear the chips. After three or four complete revolutions—enough to ensure the tap's proper alignment—free the tap from the chuck or tailstock center; then finish using a hand wrench. Do this with the work still in the chuck or taken to a bench vise.

Why turn the spindle by hand? Why not run the motor? You certainly can, but any tap smaller than #10 or M5 is too fragile to risk it, in my opinion (power-tapping in the mill versus the lathe is a different matter). If you do try power-tapping in the lathe, be sure that the drive system operates smoothly at low speeds. Some lathes are very "notchy" below 100 rpm.

For chip removal (especially in a blind hole), it may be necessary to back the tap out completely. Use a magnetized scriber point or thin wire to remove steel chips. For other materials, try using a pipe cleaner with embedded wire. If all else fails, an air hose may be the only remedy.

One final point—think *hexagon*. Thread cutting with large taps takes a lot of torque, often to the point where round work slips in the chuck, leaving unsightly scars or worse. You can try further tightening of the chuck, but overdoing it can damage the scroll and jaws. A guaranteed fix is to *use hex-section* stock to do the threading; then do the finish turning later.

2-41 DIES

Most of the dies used in the machine shop are referred to as "adjustable button dies" (Figure 2-37). "Adjustable" means they have a gap that can be expanded a few thousandths by a jacking screw to increase the pitch diameter, thus allowing a better fit with a mating screw. The range of adjustment is usually up to about 0.006" or 0.008" on the diameter (but with too much expansion, the die may no longer fit into the die holder).

Adjustable dies come in several outer diameters up to 3". The two that matter most in the small shop are 13/16" and 1", covering all threads up 1/2" and M12.

Adjustability is a mixed blessing. It's certainly true that some degree of control over pitch diameter can be helpful. On the other hand, who can remember what a die was set to when it was last put back in the drawer? A workable routine might be to set the die to minimum diameter (fully relaxed) after each use—but the downside is the need for resizing every time the die is used, plus extra wear on the jacking screw. (Some users feel so strongly about this, they keep two or three same-size dies preset to different pitch diameters.)

FIGURE 2-37 Threading dies. The adjustable "button die" at left has a nominal outer diameter of 13/16" (0.8125). In the center is another adjustable die with an OD of 1". The smaller diameter is commonly used for threads up to 5/16", the larger for threads up to 1/2". The 1" hexagonal die *at right* is hobbyist grade, nonadjustable (usually not HSS), intended for use mainly as a thread restorer.

To use dies on the lathe, you will probably need two die holders, one for each of the two common sizes (Figure 2-38). The shanks of these holders, typically 1/2", are hollowed to accommodate a couple of inches of the threaded rod as it is drawn through the die.

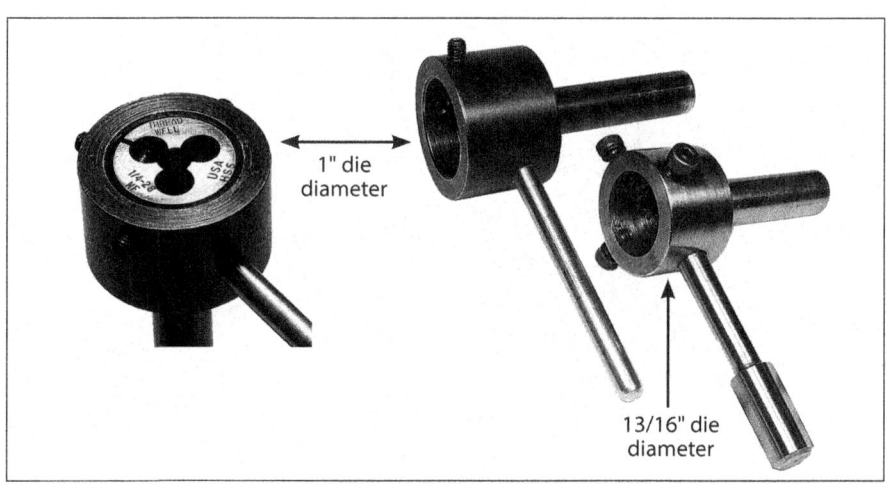

1" die diameter

13/16" die diameter

FIGURE 2-38 Die holders for the lathe. Die size and other information are always etched on the inside-tapered "starting face" of the die, *at left*.

The holders have three screws at 90° spacing. The middle screw, ideally bull-nosed, prevents rotation by engaging a dimple in the rim of the die.

The thread form inside the die is tapered, more open at the face than the back. This means that the thread should be started with the die face-out, shown at the left in Figure 2-38. It also means that for a consistent thread over the full length of the screw, you may need to finish it using a manual die stock (Figure 2-40), with the die *face-out*, the "wrong way."

A *golden rule* when die threading, on the lathe or otherwise, is to chamfer the end of the workpiece to allow easier engagement in the die's inside taper.

Die holders like those in Figure 2-38 can be held in the tailstock chuck, "slidably" snug—but not tight—with the handle stopped from rotating by hand-holding or by resting it on any suitable cross-slide surface. A better type of die holder, usually shop-made, is a two-part device (Figure 2-39): (1) a mandrel that is held *rigidly* in the tailstock chuck, and (2) a die holder that slides on the mandrel. I have two versions, one for 13/16" dies, the other for 1". The knurled grip is easier to handle—certainly for the first two or three turns—than the conventional type with only a tommy bar.

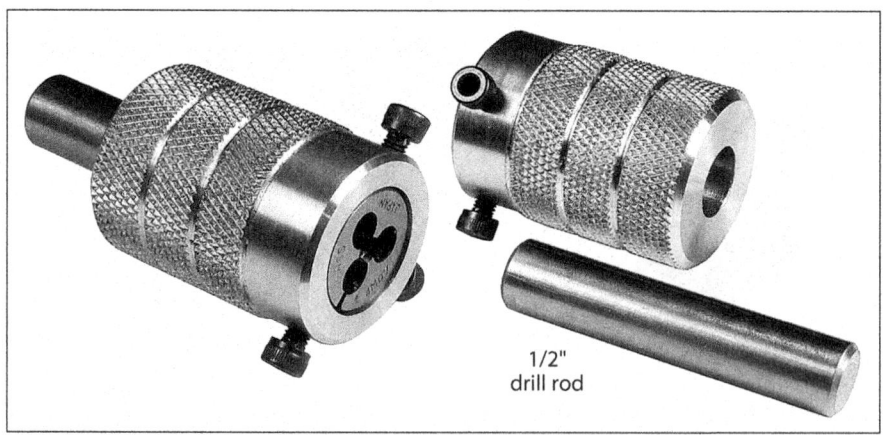

1/2"
drill rod

FIGURE 2-39 An easier-to-use die holder. Use a mild steel or aluminum bar, 1-1/4" diameter for 13/16 dies or 1-1/2" diameter for 1" dies. *Important: The cavity must be large enough to accommodate fully expanded dies.*

Before you make a holder like the one shown in Figure 2-39, check the outside diameter of your dies. The nominal diameter of 13/16 dies is 0.8125", so the holder cavity needs to be a few thousandths over that to allow for expandability. Start at say 0.815", checking a few expanded dies while the holder workpiece is still in the chuck. A corresponding starting number for 1" dies might be 1.002".

No matter what style of die holder you are using, the procedure starts with the workpiece chamfered and lubed, firmly held in the chuck. Lock the tailstock to the bed, and apply forward pressure using the tailstock handwheel. When you are sure the *die is square* to the workpiece, increase pressure while turning the spindle by hand to pull the die holder forward. Complete the threading by additional hand-turning of the spindle, adding pressure from the handwheel every few degrees of rotation. Every turn or so, back the spindle to clear the chips. Occasionally, if I am cutting a thread longer than about 3/4", I might use low-speed power instead of hand-turning (but the last couple of turns would be made without power).

The classic problem of die threading is the tendency to run the die out of square onto the workpiece, with the result that the nut (or whatever else is supposed to fit the thread) will wobble, or worst case, won't fit all. This is more than likely to happen if you are working freehand. With a hand-held die, the only way to get close to square is careful "eyeballing" and correcting during the first couple of turns of the thread. The problem largely goes away if you are working on the lathe with a die holder such as one of those shown in Figures 2-38 and 2-39, or if you are using a pusher device as shown in Figure 2-40. In some cases, especially with coarse threads on tough material, it can be helpful to *start the thread with several passes of a single-point cutting tool*, then finish using a die.

Again, think *hexagon*! Several times elsewhere, it has been pointed out that thread cutting takes a lot of torque, causing slippage in the chuck and other problems. Using hex-section stock instead of round is one solution.

The usual remarks about sets apply to dies—absent a knockout deal, don't be tempted into buying a set. For dies only, I take back what I've said

FIGURE 2-40   13/16" die stock. This can be used in place of a dedicated tailstock die holder. To hold the die square to the workpiece, use a 1/2" rod in the chuck as a pusher. Use the tailstock handwheel to apply pressure.

about HSS. For occasional use, you can get by just fine with non-HSS dies, even on steel if properly lubricated.

2-42 REAMERS

For sizing holes accurately with very little effort, nothing beats a reamer, full name "chucking reamer" (Figure 2-41). These come in two styles, straight flute as shown, and, less usually, helical (spiral). There is not much to be said about the relative merits of the two styles, but one thing is certain—if the hole to be reamed has a key slot or cross-hole, spiral flutes are preferable.

Reamers are used to enlarge precisely a hole that's a few thousandths undersized—no more than 0.005" on the diameter for holes between 3/16" and 1/2", even less for smaller holes. If you leave more material, the reamer may be pushed out of line, oversizing the bore. The other thing to bear

in mind is *spindle speed*, especially
when reaming steel—about one-
half of the speed you used for drilling, and be sure the reamer is lubed
to lengthen tool life.

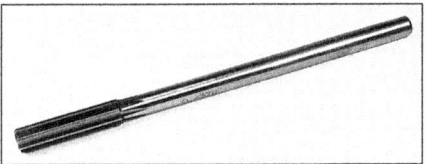

FIGURE 2-41 Chucking reamer.

Golden rules for reaming in the
lathe:

- Run the spindle *slower* than the drilling speed.
- Lube thoroughly.
- Leave only a small amount of material for reaming.
- Don't push a reamer hard against the bottom of a blind hole—it will cut oversize if you do.
- Withdraw the reamer slowly, spindle still turning.

Not knowing how fortunate I was at the time, I acquired from a retiring
toolmaker two sets of straight-flute reamers, 14 sizes each set. The reamers came in index boxes, one marked "dowel pin," the other "over/under."
Today, they are right up there on my most-used cutter list. My two sets are
listed in Table 2-2:

TABLE 2-2 Dowel Pin and Over/Under reamer sets

Nominal	**1/8** 0.125		**3/16** 0.1875		**1/4** 0.25		**5/16** 0.3125	
Dowel pin	0.123	0.1247	0.1855	0.187	0.248	0.2495	0.3105	0.312
Over/under	0.124	0.126	0.1865	0.1885	0.249	0.251	0.3115	0.3135

Nominal	**3/8** 0.375		**7/16** 0.4375		**1/2** 0.5	
Dowel pin	0.373	0.3745	0.4355	0.437	0.498	0.4995
Over/under	0.374	0.376	0.4365	0.4385	0.499	0.501

- *Dowel pin* reamers are used to size holes for dowel pins when
  you need them to be really tight-fitting (so tight *they can
  distort the work* if you don't pay attention).

- *Over/under* reamers have equally obvious applications. They are used mostly to size bearings for standard-size shafts.

Unless you are as lucky as I was, it usually makes sense to buy reamers individually as the need arises (HSS only, of course). Beware of "expanding reamers." They are mostly non-HSS, and in my experience have never delivered a usable result.

2-43 CUTTING SPEEDS FOR KNIFE TOOLS

Any attempt to get "the one answer" to the question of how fast to run a lathe always turns into a voyage of discovery, with no certain result. Aside from the ordinary problem of misinformation—doubtful data copied and recopied over the years—the fact is there are too many variables in play for a definitive, single answer.

The first two variables are workpiece diameter and workpiece material. For a given spindle speed, the larger the workpiece diameter, the faster its surface speed where it meets the cutting tool. Equally clear is that softer materials such as aluminum and brass can be cut at higher surface speeds than steel can be.

Recommended Cutting Speed

This is measured as surface feet per minute, abbreviated to SFPM. In machining terms, this is the speed of the workpiece circumference at the point where the cutting tool contacts the workpiece (see Table 2-3).

So much for the easy stuff. We also need to weigh in factors such as:

- **Material specifics.** If it's supposed to be mild steel, exactly what alloy is it? There are dozens of them, mostly simple "low-carbon" alloys, but there are a few with more exotic constituents. The same applies to stainless steel, which comes in various levels of machinability.

- **Coolant.** Are you machining dry, or with a continuous supply of cutting fluid? Water-miscible oil keeps the cutting tool cool, allowing higher cutting speeds.
- **Type of cutting tool.** Carbide tools allow cutting speeds between two and three times faster than HSS, at least in theory. Carbide tools differ, too—some are plain, others nitride-coated, etc.
- **Cutting tool condition and geometry.** How sharp is it? Are the rake and relief angles right for the job?
- **Tool wear.** What compromise between machining time, resharpening time, and replacement cost works best for you?
- **Depth of cut.** Industrial capability, 1/4" plus, or 50 thousandths of an inch in the model shop?
- **Machine weight/rigidity.** Industrial weight (2 tons plus) or model shop (less than a ton, perhaps around 500 lbs. or less)?

2-44 SPEEDS FOR DRILLING

This raises another important question. First, some definitions: in drilling operations, *spindle speed* is the rotational speed, in rpm, of the drill press spindle; or, in the case of a lathe—with a stationary drill in the tailstock chuck—it is the rotational speed of the workpiece. *Diameter* is the diameter of the drill.

The FAQ here is: *What is the relationship between SFPM and spindle speed?* There are tables of recommended SFPM values for every metal under the sun; but you won't find a table that gives you spindle speed, because you need also to specify a diameter, meaning the diameter of the drill, or of the workpiece, at the point *where it meets the cutting tool* in the lathe.

Example: We are drilling mild steel, with a suggested SFPM of 100 ft/min. Suppose the drill is 1/4" diameter, circumference = 0.25" x $\pi/12$ = 0.065'. To achieve a surface speed of 100 ft/min, we must run the drill at 100/0.065 = 1528 rpm.

Aside from the possibility that such a speed may not be available, 1528 rpm is way faster than most small shop machinists would consider reasonable. Even more unlikely would be the speed called for if the drill were smaller—say 1/8", speed 3056 rpm. This underscores the main point in all discussions to do with machining speed—namely, "the books" offer the best compromise for commercial operations, with all of the above variables optimized. Those numbers may not necessarily—nor even probably—apply in non-manufacturing situations.

The small-shop answer is to think of the calculated spindle speed as a point of departure. In practice, you simply go with the nearest available speed, certainly for early experiments. After a while, you begin to develop an instinct for what works best in your setup. Don't hesitate to differ—even quite widely—from the calculated numbers. You will know when the spindle speed is too high—squealing, chattering, tool and chips overheated; and when the speed is too slow—"hogging," thick chips, tool jumping, stress on the workpiece.

There's an easy way to estimate the spindle speed in a couple of seconds: Multiply the SFPM value by four,* then divide the result by the drill (or workpiece) diameter. (*4 is an approximation; the true value is $12/\pi = 3.8197$.)

Spindle speed $=$ SFPM x 4 $\div$ Diameter

Simply put: If the drill diameter (or the workpiece) is 1", the "book" speed in rpm is the same as SFPM x 4.

Example (1) mild steel: Suggested SFPM $=$ 100, SFPM x 4 $=$ 400, Diameter 1/2", speed 800 rpm.

Example (2) unhardened tool steel: Suggested SFPM $=$ 50, SFPM x 4 $=$ 200, Diameter 5/16", speed 640 rpm.

Table 2-3 lists suggested SFPM values for six metals commonly machined in the small shop, together with calculated spindle speeds for diameters up to 1". The table is based on rough averages of the SFPM numbers published in *Machinery's Handbook*. If using carbide cutting tools, consider *doubling* the listed spindle speed.

TABLE 2-3   HSS cutting speeds in SFPM and spindle speeds in rpm

	Cutting Speed SFPM	SFPM x 4	Suggested Spindle Speed (rpm)					
Drill or workpiece diameter			1/8" 0.125	1/4" 0.25	3/8" 0.375	1/2" 0.5	3/4" 0.75	1" 1.0
Stainless steel	40	160	1280	640	425	320	215	160
High carbon tool steel	50	200	1600	800	530	400	265	200
Cast iron	70	280	2240	1120	750	560	375	280
Mild steel	100	400	3200	1600	1070	800	535	400
Brass	200	800	6400	3200	2140	1600	1070	800
Aluminum	300+	1200	9600	4800	3200	2400	1600	1200

Speeds not usually available on general-purpose machines.

2-45 CUTTING SPEED AND TOOL LIFE

Tool life is affected *mostly* by cutting speed, followed by feed rate—how fast the tool traverses along the work. The factor with the least effect on tool life is depth of cut, which has mostly to do with the power available and the rigidity of the workpiece and the cutting tool.

CHAPTER 3

Add-Ons to Make Life Easier

CONTENTS AT A GLANCE

3-1 IS THE COMPOUND REALLY NECESSARY?

What an odd question! Surely it must be—every metalworking lathe comes with a compound (Section 1-10, Figure 1-9).

Yes, it *is* necessary, but not as often as you might think. There are two operations for which it is a definite requirement: (1) taper turning of short spindles, up to about 3" long or more (the range of compound travel), and (2) screw cutting, for which the compound is usually at 29° or 30° relative to the cross-slide axis.

For *everything else*, though, the compound can be replaced by a solid block of steel (Figure 3-1).

FIGURE 3-1   A steel block is more rigid than the compound.

I first tried this many years ago when I discovered that the compound on my Emco Compact 8 wasn't rigid enough for serious depths of cut. I also noticed that backlash in the compound occasionally caused minor errors if I wasn't paying attention. Now, three lathes later, my 12" lathe has the original 3/4"-thick block, now sitting on a second 3/4"-thick block to make up the difference in center height. For screw cutting, I use a shop-made 29° retracting toolholder (Section 3-3 and Chapter 11).

Using a solid block instead of the compound is strictly a personal preference. There is absolutely no need to do this, but you might appreciate some of its benefits, especially with lightweight lathes. It is mentioned here only as a possible enhancement.

3-2 ADD GUIDELINES FOR SPECIFIC JOBS

Aside from its greatly improved rigidity, the steel block has one other significant benefit—it can be marked with fixed guidelines to position the tool post, which is assumed here to be a QCTP (Figure 3-2), not a 4-way turret (Chapter 1). For instance, a line at 90° can be used for cutting off and grooving (Figure 3-3). (You could mark a compound in the same way, but that works only for one specific compound setting.) Figure 3-4 shows the tool post turned a few degrees counterclockwise.

FIGURE 3-2   QCTP toolholder with RH knife tool (carbide insert).

Another useful guideline helps in finishing the workpiece with a light chamfer. You could do this with a file or an angled cutting tool, but it is more predictable and nicer looking if instead you use a left-hand knife tool with an indexable 60-60-60 insert. Set the tool post on the 15° line (Figure 3-5).

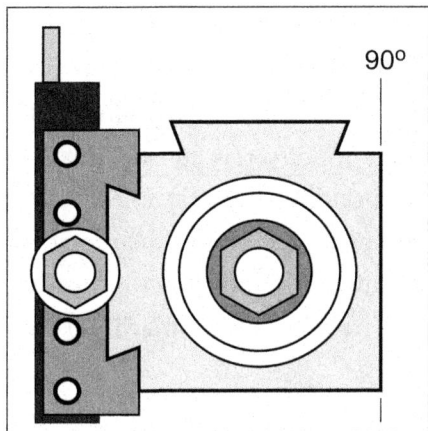

FIGURE 3-3 QCTP tool post set at 90° for cutting off. The 90° line can also be used to estimate a clockwise offset of a few degrees for a facing operation and for cutting a diameter down to a square shoulder.

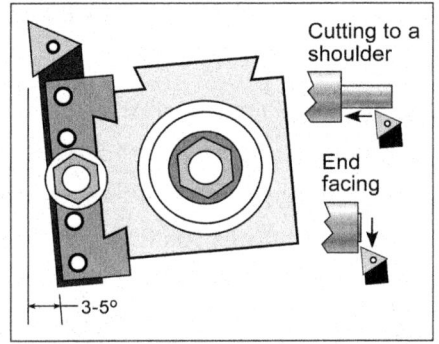

FIGURE 3-4 QCTP tool post turned a few degrees counterclockwise. With many RH knife tools, this allows both end-facing and traversing (diameter-reducing) cuts without further adjustment. The angle is not critical and can usually be estimated.

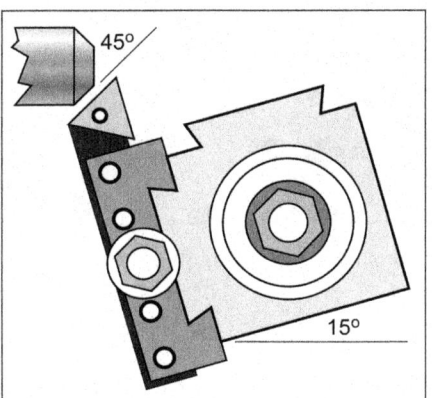

FIGURE 3-5 QCTP tool post turned 15° for chamfering.

3-3 A NEW WAY OF SCREW CUTTING

Finally, you can add 29° lines for screw cutting using a special QCTP tool-holder with its own lead screw, essentially a fixed-angle "compound" with a built-in retractor (Figure 3-6). If you do a lot of screw cutting, this is a great time-saver. Setup is a breeze: You can set the necessary infeed angle instantly, simply by rotating the QCTP tool post—a lot easier than rotating the compound. You won't find the screw cutting 29° toolholder in the catalogs, but it is an interesting shop project (needs a milling machine). See Chapter 11 for more.

FIGURE 3-6   Special 29° toolholder for screw cutting.

3-4 ROUGH-CUTTING METAL WITH A BANDSAW

Every project calls for some means of rough-cutting metal to a size that can be handled in the lathe. You can do this the old-fashioned way, using a hacksaw, but this takes a lot of effort and time. Pretty soon, you will be looking for a power saw. Most small-shop machinists are happy with the 4" x 6" metal-cutting bandsaw sold by importers such as Grizzly and Harbor Freight (Figure 3-7).

FIGURE 3-7   A 4" x 6" bandsaw.

This is a design that has been around for years. It cuts plastics and most metals, including unhardened steel, in sizes up to about 4-1/2" round and rectangular sections up to 4" x 6".

The blade runs in an oval-shaped frame that swings down onto the workpiece by gravity, counterbalanced by an adjustable spring. Cutting rate is determined by the combination of blade speed (choice of three) and downfeed pressure. For initial trials, start with low blade pressure (high force from the spring); then work up to the most efficient setting for the material in process.

The 4" x 6" bandsaw usually ships with a 14-TPI (teeth per inch) blade, satisfactory for most work on unhardened metals. Replacement blades are available from McMaster Carr and other suppliers. The least expensive general-purpose blade is carbon spring steel, good for all unhardened materials. Usually more expensive—but worth the extra if you cut hardened/welded steel—are bimetal blades fabricated from high-speed steel teeth welded to carbon spring steel.

Blade tension is important, but you get no help from the instructions. On larger commercial saws, the tension can be metered precisely with an external gauge, but the 4" x 6" saw is not in that league. The best you can do is experiment, putting up with the fact that the blade frequently falls off its wheels. Blade tension too low, the blade falls off: too high, the blade falls off. This is something you get used to, and quite soon you have the reinstallation process down to a fine art. (If you look at the geometry, it should not be surprising that the blades fall off—at each end of the very short frame, the blade has to go continuously through a directional change of 45° or so.)

I found over many years that a 4" x 6" saw works best with the blade tight enough to produce a musical note when twanged, not a dull thud. That's about as specific as it gets. Lubricant on the blade helps preserve the blade, and it also increases cutting speed—coat both sides of the blade, while it's running, with a waxy lubricant stick. Alternatively, especially for steel, brush on your usual cutting/tapping fluid every 1/2 minute or so.

If you have the floor space and budget for it, consider a higher-capacity bandsaw such as shown in Figure 3-8. This saw cuts much faster and more consistently. It comes with a hydraulic feed controller that's a lot easier to adjust than the counterbalance spring on the more basic 4" x 6". Capacity of the model shown is 7" round, rectangular sections up to about 7" x 10". Sometimes included with saws of this size is a coolant system with a recirculating pump housed in a 4-gallon tank—essential for high-volume steel cutting but a bit over the top for the small shop; instead, use a spray bottle with Swisslube or other coolant (Section 4-10).

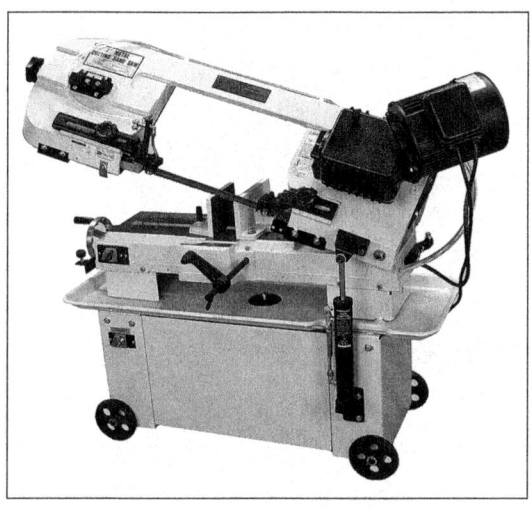

FIGURE 3-8   A 7" x 10" bandsaw.

3-5 BLOCK SQUARING THE EASY WAY

If you don't have a mill—or even if you do—squaring the surfaces of metal stock is a time-consuming business. A workaround I have used for years is an 8" disk sander made from a salvaged Delta-style belt/disk sander (Figure 3-9). The table it came with was just about usable, but I replaced mine with a beefier steel version that was carefully squared against the disk. I made two fixed-angle slides to go with it, 90° and 45° (for mitering), firmly pinned to 3/4" x 1/4" rectangular bars.

The disk sander is a great time-saver, used to prepare stock on at least 50% of my projects. Highly recommended. It may be that commercially made disk sanders are accurate and rigid enough to do what mine does—or can be shimmed and tweaked to take care of minor discrepancies.

FIGURE 3-9 Modified disk sander.

3-6 SPINDLE INDEXING

Once you have a lathe up and running. it doesn't take long before you wish you could click the spindle around by exact angular amounts. Maybe you need to drill a series of through holes evenly around a flange, or radial holes for spokes on a wheel. First, we need to look at ways of indexing the spindle.

The answer given in most lathe books over the past 100 years is to attach to the spindle one of the external change gears that came with the lathe. A 60-tooth gear, for example, gives these divisions: 2, 3, 4, 5, 6, 10, 12, 15, and 30. Since most lathe spindles are hollow, the easiest way to attach the gear is with an expanding mandrel. For more on indexing, see Chapter 9.

With the gear firmly attached to the spindle, all you need is some form of detent, shaped to fit between the gear teeth. Typically, the detent will be held in place by a spring that allows it to be lifted clear for indexing to the next position. For really solid indexing, the detent must conform closely to the teeth, and the spring has to be surprisingly heavy.

Over time, detent and spring issues (plus the 60-division limitation) prompted me to replace the gear with a custom dividing disk and heavy-duty indexing pointer, shown in Figure 3-10. The complete assembly on an earlier lathe is shown in Figure 3-11.

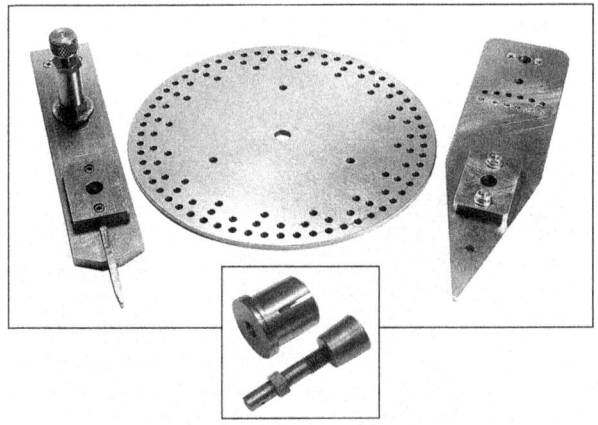

FIGURE 3-10 Indexing components. The indexing pointer on the right is a vernier that allows 1° increments, like a milling machine spin indexer. An expanding mandrel, shown inset, is a reliable way to attach an indexing plate (or 60-tooth gear).

FIGURE 3-11 Spindle indexer installation.

3-7 FILING FLATS ON A WORKPIECE

Time and again, there's a need for flats on a circular piece in the chuck. Perhaps you need a pair of flats for an open-ended wrench, or maybe a hex head on a custom screw. Conventionally, you would cut off the part, then transfer it to the milling machine (assuming you can find a workable way to hold it). An easier way, if you have a spindle indexer and a simple filing rest (Figure 3-12), is to file the flats, then cut off the finished part. This might sound a little offbeat, but it works beautifully.

FIGURE 3-12   Filing rest installed in a standard QCTP toolholder.

There is more information on the filing rest in Chapter 9. There's not much to it—two 3/4" rollers, preferably case-hardened, with flanges on both sides to keep the file on track. They run on shoulder bolts spaced 3" apart on a 1/2"-square steel bar. The filing rest is clamped in a QCTP toolholder in the usual way. It can be positioned in front of the workpiece (Figure 3-13) or straddling it for better control (Figure 3-14). Depth control is a breeze. Starting with the file just grazing the workpiece, cut a full circle of flats. Measure the result, and then lower the toolholder in a series of steps, cutting a full circle at each step.

FIGURE 3-13   Filing rest used to cut hexagon flats on a workpiece.

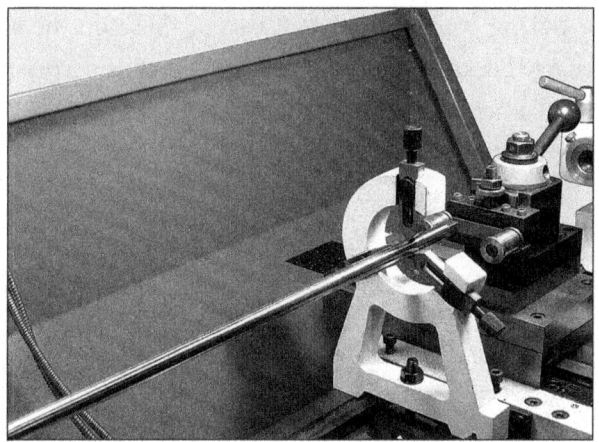

FIGURE 3-14   Use a steady rest to file flats on a long workpiece.

3-8 DRILLING FROM THE TOOL POST

Suppose you need a number of evenly spaced holes in a flange. Conventionally, you would take the turned flange from the lathe to a milling machine, then run a PCD (pitch circle diameter) routine. To do that, you would need a digital readout, or you could use a rotary table instead. *Could we drill it on the lathe instead?* Yes, if you have a spindle indexer and a tool-post drill (Figure 3-15).

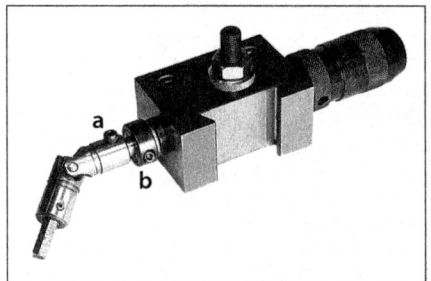

FIGURE 3-15   Tool-post drill. The socket head screw (a) and clamp collar (b) allow the assembly to be reversed for either longitudinal or radial drilling.

FIGURE 3-16   QCTP boring bar holder.

The starting point for the drill was a 100 Series "heavy-duty" QCTP 3/4"-diameter boring bar holder. I had two of these on the shelf, gathering dust for years (Figure 3-16). The tool-post drill is surprisingly easy to make, and a great asset to have on hand. Aside from anything else, it allows off-axis drilling and tapping with no danger of misalignment. See Chapter 9 for details.

FIGURE 3-17   Tool-post drill setup for longitudinal drilling.

FIGURE 3-18   Tool-post drill setup for radial drilling.

3-9 TOOL-POST GRINDER

This has nothing to do with indexing the spindle. It is mentioned here because of its similarity to the tool-post drill, at least in its construction (Figure 3-19).

Like the tool-post drill, the grinder started out as a QCTP boring bar holder—the second one I could never find a use for. That was until the day a number of previously made parts needed to be modified using an ancient Dremel flex-shaft rotary tool. Hand-holding might have done the job, but not with the desired precision. It turned out that

FIGURE 3-19   Tool-post grinder. The arrow points to the flex-shaft coupling.

the chuck assembly at the active end of the flex shaft could be made to fit nicely in the boring bar holder with only a minimal amount of tinkering.

The budget tool-post grinder enabled a number of previously impossible tasks, including operations on already hardened workpieces (Figure

3-20). Cutting off in the lathe turned out to be easy and accurate—so also is grinding a small amount off the shank of HSS cutters.

Having said that, tool-post grinding is not for me unless I really have to do it. Even with the most careful preparation and masking off, abrasive dust gets everywhere. Nevertheless, it's good to have this little grinder on hand for times when there's absolutely no alternative.

FIGURE 3-20 Cutting a groove in a hardened shaft.

3-10 TAILSTOCK DRILL PRESS

This is a great time-saver when drilling small, deep holes, especially when there's a need for frequent withdrawal for oiling and cleaning. Additionally, lever operation gives a degree of sensitivity that's simply not possible with the standard lead-screw feed. For that same reason, the tailstock drill press (Figure 3-21) is also very useful for feeding in small, fragile taps.

The starting point for the tailstock drill press is a machinable MT3 to MT1 adapter sleeve (aka "taper extension socket"), drilled for a 1/2" drill-rod quill. This design works for other tapers from MT2 and up (MT3 is the taper of my lathe's tailstock). See Chapter 10 for details.

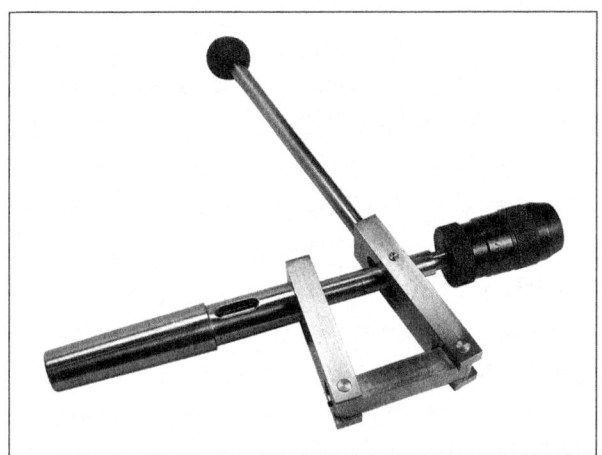

FIGURE 3-21   Tailstock drill press.

FIGURE 3-22   Tailstock drill press in use.

3-11 CENTER HEIGHT GAUGE

The gauge in Figure 3-23 sits on the cross-slide, with the *upper surface of the blade* at the exact center height of the lathe. This example is unnecessarily fancy, but it works well. I have found it easier to adjust tool height with a flat-topped blade like this than with other, more pointy, gauges.

FIGURE 3-23   Center height gauge.

3-12 AN INDISPENSABLE FINISHING TOOL

How many times have you drilled holes on a finished surface and wish you had a way of removing the burrs? A file doesn't do it, because that's fatal to the finished surface. The easy answer, one that works every time, is a pin vise with a countersink bit (Figure 3-24). I have two pin vises, both 1/4" capacity, one with a countersink, the other available for hole clean-out using a drill or reamer.

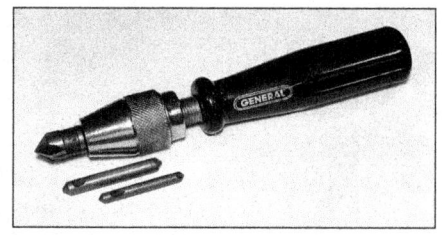

FIGURE 3-24   Pin vise with various countersinks.

3-13 WORKPIECE CLEAN-UP

I'm assuming here you don't use an air hose to clean off machines and workpieces. Commercial machinists do that in the ordinary course of business, but they are not as troubled as you might be by the effect that can have on the walls of the shop, etc. So we almost always use the less-messy shop vac instead. That works for machines and general clean-up but usually does nothing for the workpiece.

What causes the most frustration in cleaning up the workpiece is usually small-diameter holes, especially blind threaded holes. Try pipe cleaners of the "scratchy" sort, containing brass wire. On steel, use a thin *magnetized*

scriber (shown in Figure 3-25) or a
dedicated length of drill rod (rub it
on the surface of a strong magnet).

3-14 GENERAL CLEAN-UP
IN THE SHOP

This is something else you don't
want to use an air hose for.

The following might strike you
as really offbeat, but it can be a

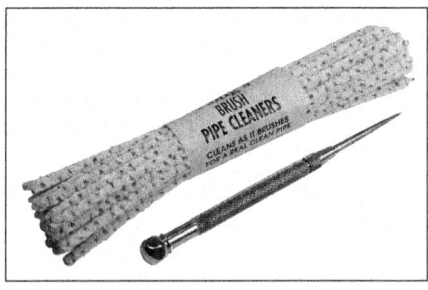

FIGURE 3-25   Two must-have items for
clean-up.

minor game-changer—no more hauling the shop vac around, tripping over
cables and hoses. The setup in my case is nothing more than a hole in the
wall at the same height as the inlet port of an ancient shop vac with a standard 2-1/2" hose. The hose, all 21 feet of it, is inside the shop (Figure 3-26).

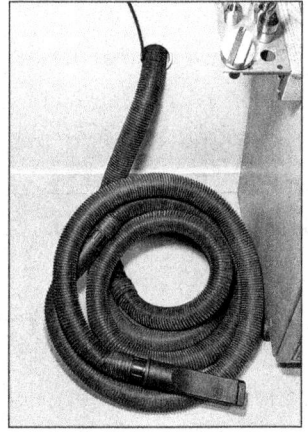

FIGURE 3-26   Stationary shop vacuum. The power cord runs through the wall to the
switched outlet inside the shop.

The vac sits outside, stationary, in a storage area. Power to the vac is
from a switched outlet in the shop. When not in use, the hose is simply
coiled up on the floor. At one time, the hose was more elegantly coiled on a
wall bracket, but the piled-up scheme works better in practice.

This has been a big improvement over previous setups using two or more smaller shop vacs with 1-1/4" hoses. They work passably if cleaned regularly, but clog easily on lengthy chips—much more so than the vac now in use. The downside of any machine shop vac is that oil accumulates on every inner surface. Hoses and filter are cleanable using hot water and detergent, but you might want to consider replacing them every year or so instead.

3-15 TWO MORE MUST-HAVES

Finally, two disposable items—flux brushes and paintbrushes (Figure 3-27)—you really can throw away without too much guilt (but you don't have to; instead, you can wash and reuse them for years). I use flux brushes to lubricate machine ways and also to apply cutting oil.

Disposable paintbrushes I keep on the bench, and on every machine. They are the handiest means I know of for cleaning off drills and cutters before returning them to the drawer—something neglected for years until it began to dawn on me that this really is an overall time-saver.

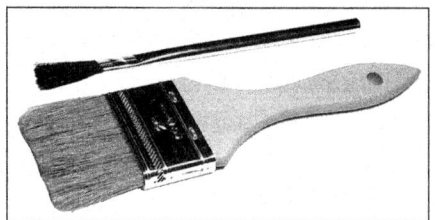

FIGURE 3-27 Multiuse flux brush and paintbrush, both disposable.

SPEAKING OF CLEAN-UP . . .

Tool-post grinding (Section 3-9) is not the only source of abrasive dust. More routinely, most machinists use emery paper, Scotch-Brite, and other abrasives to work on surface finish, etc.

The golden rule is to protect *every surface* on the lathe, including the chuck, with scrap paper—fit a large circular paper washer between the workpiece and the chuck jaws.

CHAPTER 4

Things I Wish Someone Had Told Me

To be honest, if I include any number of patient instructors, plus the stack of reference material and magazines accumulated over the years, I was "told" most of these things many times over. In other words, there have been dozens of contributors to this chapter. Aside from that, some of my own answers to lathe problems are in the preceding chapters. For more answers, read on. I hope you will find this a useful summary.

CONTENTS AT A GLANCE

4-1 HOW TO LEVEL THE LATHE

The objective here is to ensure that the lathe bed is in the same state as it was, one hopes, when the lathe was manufactured—level from end to end along the bed and from front to back. In other words, no warping.

Start by installing leveling mounts under the lathe cabinet or table. Make sure they are all properly weight bearing, firmly in contact with the floor. Check and adjust the level from end to end using a precision machinist's level, if available. If not, use the most reliable level on hand. Check and adjust across the bed using a matched pair of spacers such as 1-2-3 blocks to clear the Vee tenons. Alternatively, check for level on the finished surface of the cross-slide as the carriage is traversed from end to end (Figure 4-1). See Chapter 6 for more on machinist's levels.

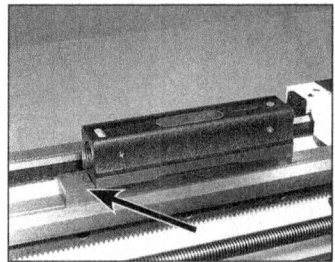

FIGURE 4-1   Check the level in both axes. Check the level of the cross-slide at the headstock and tailstock ends; if there is no twisting of the bed, the indication should be the same. To check the left-to-right axis, rest the level on a precision-ground bar, indicated by the arrow in the left-hand figure.

You can save a lot of effort and frustration if the leveling mounts can be adjusted from above, as shown in Figure 4-2.

FIGURE 4-2   Typical top-adjusting leveling mount.

4-2 HOW TO DEAL WITH UNEXPECTED TAPERS

The most important attribute of a lathe that is properly set up is its ability to "machine parallel," to cut a cylinder of uniform diameter over its entire length. In other words, no taper.

Leveling of the lathe is a part of this. Equally important is the alignment of the center-to-center axis relative to the lathe bed, as seen *from above*. (Vertical alignment is not nearly as critical—rarely a cause of taper unless the lathe is damaged or badly worn.)

4-3 ALIGNING THE TAILSTOCK

Practically all lathes come with some means of offsetting the tailstock, usually for taper turning. For routine operations, the offset must be zero (Figure 4-3).

The scale usually provided on the tailstock is not reliable for precision work—think of it as only a starting point.

What follows are two methods for aligning centers, one quick and easy, the other more precise.

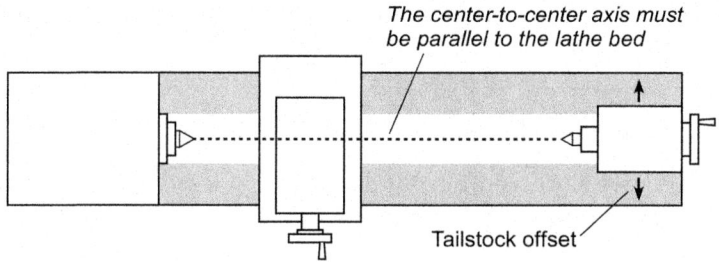

The center-to-center axis must be parallel to the lathe bed

Tailstock offset

FIGURE 4-3 Center-to-center axis.

4-4 TAILSTOCK ALIGNMENT—QUICK METHOD

This method works only if the centers are in new condition, sharp and clean.

1. Carefully clean the taper sockets and the tapers themselves. Install the tapers.
2. Move the carriage left as far as it will go; then slide the tailstock to the left.
3. Lock the tailstock (this is important—unlocked versus locked can mean an offset of several thousandths).
4. Advance the tailstock quill to bring the centers together.
5. Place a scrap of hard shim stock or an old-style double-edge razor blade between the centers (Figure 4-4).
6. Advance the tailstock quill to trap the blade; then lock the quill. If the centers are aligned, the blade will point squarely front to back. If not, adjust the tailstock offset by a series of very small adjustments.

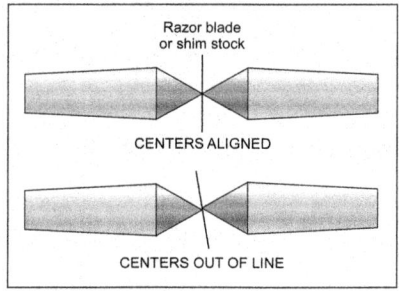

Razor blade
or shim stock

CENTERS ALIGNED

CENTERS OUT OF LINE

FIGURE 4-4 Quick alignment check. Alignment viewed from above is much more important than the view at eye level, front to back.

7. Check the blade alignment at various extensions of the quill. There should be no appreciable variation.

4-5 TAILSTOCK ALIGNMENT—PRECISE METHOD

This method uses a precision-ground steel rod at least 10" long. Look for a 3/4" or 1" drill rod with a diameter tolerance of ± 0.001" or less.

Straightness and uniform diameter are both important (but absolute diameter of the rod is not). For another way to check tailstock alignment, see the cut-and-try method described in Section 4-7.

Prepare the test rod:

1. Set the rod in a collet chuck, or independent 4-jaw chuck, with the outer end about 1/2" clear of the chuck.
2. Use a dial indicator to check for runout. If using a 4 jaw, adjust as necessary for minimum TIR (aim for 0.0005" or less).
3. Center-drill the end of the ground rod, shown in Figure 4-5.
4. Reverse the rod, readjust for minimum TIR, and then drill the other end.

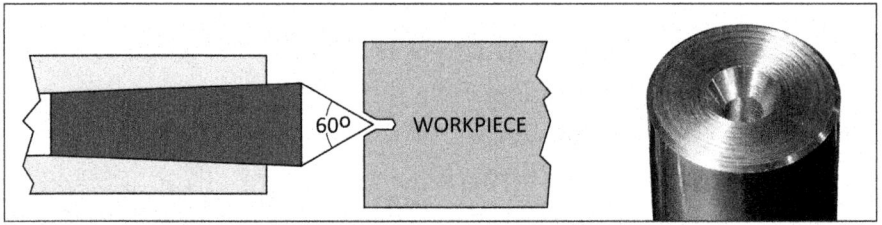

FIGURE 4-5 Center drilled test rod.

Install the test rod:

1. Set the test rod snugly between centers, as shown in Figure 4-6. Lock the tailstock.
2. Set a dial indicator on the cross-slide (to eliminate vertical error, use a flat disk contact point; if not available, machine a cap to fit over the contact point you have on hand.)

3. Starting at location 1 in Figure 4-6, note which way the pointer rotates when the cross-slide is moved inward. In this diagram, the pointer is assumed to turn clockwise as the cross-slide moves in.

4. Preload the indicator by a few thousandths; then traverse the saddle from end to end. In a perfect setup, the pointer will not move at all.

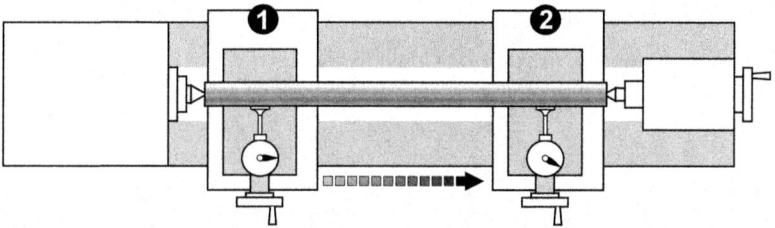

FIGURE 4-6   Test rod between centers.

If the pointer turns clockwise as you go toward the tailstock, as shown in Figure 4-6, the tailstock is biased to the front. This will cause the lathe to cut a tapered workpiece with the larger diameter at the headstock end. Correct this by a series of *very small* adjustments to the tailstock offset—see the box "Think Before Adjusting the Tailstock Offset!"

THINK BEFORE ADJUSTING THE TAILSTOCK OFFSET!

The golden rule here is this: *Don't adjust the tailstock unless you really have to.* Adjusting for true-parallel turning between centers takes a lot of iterations and is supersensitive. *Example:* If the offset screws are M6 socket heads with 1-mm pitch threads, 0.04" per revolution, it takes a hardly noticeable 9° turn of the hex key to shift the tailstock 0.001". It's even more sensitive if the screws are M8 x 1.25, in which case a turn of only 7° moves the tailstock

0.001". Bear in mind that this is a *radial shift*. The *diameter* change at the tailstock end will be 0.002".

Also note that it's easy to get the sense of the adjustment wrong, leading to the question: *When you rotate say the front offset screw clockwise, which way does the tailstock move—to the back or to the front?* The answer—it depends. If the screw is threaded into the tailstock base, *pushing* against the upper casting, it will move backward. If instead the screw is threaded into a sleeve in the upper casting, it will be *pulled* forward.

One more point: Before adjusting the offset screws, it is necessary on most lathes to release the tailstock locking lever.

Another important question has to do with headstock/spindle alignment relative to the lathe bed. For turning *between centers* this hardly matters at all; the headstock can be wildly out of square (Figure 4-7), but the lathe will still machine parallel if the centers have been aligned as previously described.

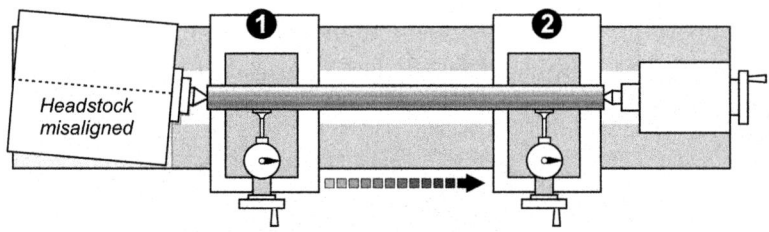

FIGURE 4-7 Misalignment of the headstock has practically no effect on turning between centers.

4-6 WHEN HEADSTOCK ALIGNMENT REALLY MATTERS

Headstock alignment isn't really a factor when turning between centers, but it's critical when the workpiece is held in a chuck or a collet, with the tailstock disengaged—often about 90% of the workload in a typical model shop. Assuming no appreciable deflection of the workpiece (too thin, too far from the chuck), taper problems in a chuck/collet setup are due to misalignment of the spindle axis relative to the lathe bed.

On some machines, this is correctable by realigning the headstock, assuming that adjusting screws have been provided in the usual places (Figure 4-8). If there is no provision for realignment, it *may* be correctable simply by loosening and retightening the headstock attachment screws.

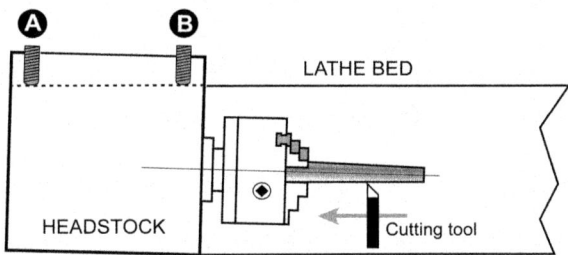

FIGURE 4-8 Misaligned headstock. In this illustration, the workpiece diameter increases as the cutting tool moves toward the chuck. Correct this by screwing in B a fraction of a turn to rotate the headstock counterclockwise, moving the workpiece away from the tool. Screw in A if the taper is in the other direction, thinner toward the chuck.

Misalignment of the spindle by even the smallest fraction of a degree causes a very measurable taper, even over short lengths of material. For example, a misalignment as small as *one-hundredth of a degree* will give a taper of 0.001" in 3". If the headstock is 10" long, this would be corrected by tapping one end of the headstock forward or back by as little as 0.002", a tiny amount even if jacking screws are provided. What this amounts to is that headstock adjustment is a *highly sensitive, iterative procedure* that should not be attempted casually.

4-7 CHECKING ALIGNMENT BY A "CUT-AND-TRY" PROCEDURE

This is in two parts: (1) headstock alignment, referring to Figure 4-8 (shown earlier), and (2) tailstock alignment, referring to Figure 4-6 (shown earlier). For both procedures, install a right-hand knife tool with a *zero degree* side cutting edge angle.

Why this? Because a cutting edge that's not 90° to the lathe axis pulls or pushes on the workpiece, causing unexpected dimensional changes.

(1) Headstock Alignment

With the *cross-slide locked* before each cutting pass, make a series of cut-and-try passes on scrap material. If the workpiece is thinner at the tailstock end, the headstock needs to be pivoted away from the tool and vice versa (Figure 4-8). This assumes your lathe's headstock alignment is adjustable—many are not—so do the best you can by careful leveling. In some cases, it's possible to achieve better alignment simply by loosening, then re-tightening the screws that secure the headstock to the lathe bed.

(2) Tailstock Alignment

A requirement for this test is a means of turning a length of scrap material between the centers. If your lathe came with a slotted driver disc (unlikely), attach it to the spindle, and attach a lathe dog to the workpiece (Section 1-16). If not, improvise something on the lines of Figure 4-9 (shown below), using a faceplate and shop-made dog.

The tailstock alignment test is based on checking the relative diameters of two turned sections, shown respectively in Figure 4-9 (A) and 4-10 (B), as far apart as possible on the workpiece. Set a stop on the lathe bed to keep the saddle, etc., clear of the dog.

Check that the knife tool is clear to cut in both locations, A and B. If available, use powered saddle feed for better surface finish and consistent diameter.

1. Center-drill both ends of the workpiece (see Section 4-5).
2. Install the workpiece between centers, with the dog engaged in the driver disc or faceplate.
3. With the tailstock locked, run the tailstock quill forward to locate the center firmly in the workpiece; then lock the quill.
4. Position the cutting tool at A; then advance the cross-slide by hand to cut fully through the skin of the workpiece, exposing

FIGURE 4-9 Improvised workpiece driver. The counterweight cuts down on wobble.

FIGURE 4-10 Tailstock with live center.

a fully machined circumference. Note the cross-slide dial indicator reading at this point.

5. Run the saddle left to cut shoulder A wide enough for micrometer measurement.

6. Back out the tool; then move the saddle right to prepare to cut B.

7. Advance the cross-slide to exactly the same dial reading as in step 4; then cut section B.

8. Measure diameters A and B. If A is smaller than B, the tail-stock needs to be offset to the back, by an amount equal to *half the difference in diameters*. If A is larger than B, then vice versa—bring the tailstock forward.

9. Cut and try at A and B again to test the adjustment just made. Readjust if necessary.

HOW HARD IS THAT TOOL?

If you have ever wondered why it is that a file can cut metal off a wood scraper, or even a hardened screw, this may help. These numbers are fairly arbitrary and highly variable, depending on manufacturers' preferences (and quality, which is all over the map—literally).

TABLE 4-1   Typical Rockwell C-Scale hardness values

Product	Hardness	Product	Hardness
Micro-grain carbide	75	Locking pliers	55
High-speed steel HSS	63–65	Hex wrenches, screwdrivers	50–54
Files	65	Machinist scales (steel rules)	50
Case hardened dowel pins	60	Axes, hatchets	45–55
Single-edge razor blades	58	Socket head cap screws	37–40
Wood chisels, plane blades	55–66	Wood scrapers	48–51
Hobby knife blades	57–59	Threading tap	60
Cold chisel	55	Hammer face	45–50

4-8 WHERE TO FIND ACCESSORIES AND MATERIALS

This is a question that comes up in practically every conversation having to do with the model shop.

There are not as many machine tool suppliers as there were 20 years ago, but those that remain are very active. *Four examples:*

1. Precision Matthews, PA (lathes from 10" x 22" to 14" x 40", bench and knee mills, table widths from 27" to 54")
2. Little Machine Shop, CA (small lathes and mills, many accessories)
3. Wholesale Tool, MI (formerly Victor Machinery, NY—wide range of accessories, measuring equipment, and cutting tools)
4. Grizzly Industrial WA and MO (wide range of wood and metalworking machines and accessories)

For peace of mind, you need to be comfortable with your supplier's product support. *Will you be able to talk to an actual user of the machine?*

For materials, there is usually no option other than buying online, because most local metals stockists have disappeared. Online suppliers offer a good range of metals in the usual choices of alloys, shapes, and sizes. The downside of online buying is the cost of freight, which can be startling—so try to get a quote.

One big-name supplier you can rely on to stock just about everything to do with machines is MSC. Another reliable supplier is McMaster Carr, for many years the go-to source for every conceivable item of hardware, including metric.

You can also go to McMaster Carr for a very wide range of materials, available overnight if you need it. *One caveat*: The company does not (or used not to) quote freight costs up front; you simply have to pay what's added to the bill for UPS, etc.

4-9 DO I NEED CUTTING OIL?

In most cases, *no* if you are machining plastics or taking light cuts on most metals, including general-purpose low-carbon steel (less than 0.2% of carbon). This applies also to cast iron (about 4% carbon), which machines quite differently from other metals—expect powder instead of chips at the cutting tool. Things change as you become more adventurous, taking deeper cuts to save time. Still, no need for cutting oil on brass and aluminum, but it may give you better results on steel. On steel, one thing it will definitely do is save wear on the cutting edge. It may also improve surface finish.

For first experiments try one of the off-the-shelf cutting oils that come in little bottles (Tapmagic, Tapfree, Cool Tool, Relton, etc.). Squirt a little oil into a disposable plastic cup, and apply using a brush of the sort used for solder flux (also disposable but in this application, it can last forever).

4-10 INDUSTRIAL CUTTING FLUIDS

In some instances, you will get better results with the cutting fluid used in commercial machining, on practically every metal including aluminum. This is a *water-miscible* cutting fluid such as Blaser Swisslube. You might be horrified to think of spraying a thin white fluid that's mostly water on your favorite machine, but machine shops do that all day, every day, on their CNC machines.

Water-based cutting fluids perform two important functions:

1. They assist cutting action, just like Tapmagic, etc., often helpful when parting off and performing other troublesome operations.

2. They cool the cutting tool and workpiece. You won't hear much about water-based fluids in model engineering books, because the concentrate is sold only in expensive multi-gallon containers. Another factor to be aware of is the mix ratio, which has to be precise—something on the order of

1 part concentrate to 10 parts water. If the mixture is too weak, it can cause rusting and other undesirable effects. So instead of buying a huge amount of concentrate and mixing your own, see if you can buy a gallon or two of premixed fluid from a local machine shop, enough for the lathe's built-in coolant system if it has one. Be sure your supplier uses a *refractometer* to check concentration—this needs to be spot-on.

4-11 SPRAY BOTTLE INSTEAD OF A COOLANT PUMP

If you don't have a coolant system, use the same fluid in a plastic spray bottle—literally, just like the garden variety. Most shops seem happy to sell premix in small quantities, so don't overbuy. If you keep the premix out of the light in the coolant tank, or in a carefully cleaned 1-gallon bottle (e.g., for distilled water), it will be good for a year or more. As it deteriorates beyond the point of usability, you will notice discoloration and oil-water separation, possibly even a bad odor. Evaporate unusable fluid in an open tank; then dispose of the residue like you would any other waste oil (maybe take it to a local auto repair shop).

4-12 LUBING TAPS AND DIES

A word of caution here: For thread cutting with taps and dies, water-based cutting fluids don't work well at model shop speeds (even though they are used routinely for thread cutting on CNC machines). The better solution is usually one of the "little bottle" cutting oils mentioned earlier. These can be applied drop by drop to the precise location, and are thick enough to stay on the spot.

4-13 THREAD CUTTING ON STEEL

For taps and dies on steel, stainless and other difficult metals, I use Castrol Moly Dee. It's expensive but definitely superior to anything else out there.

The smallest amount you can buy is 16 ounces, good for several lifetimes in the model shop. I bought 3 or 4 ounces of it from a local machine shop years ago, and there's still a lot of it left. Highly recommended.

4-14 CHOOSING THE RIGHT KIND OF STEEL

There are two main classes of steel—hot rolled and cold formed (or cold rolled). Hot rolled is the product found in steel construction shops where the material is typically flame-cut and finished—if at all—by grinding. The most common hot-rolled steel alloy is A36. It is machinable—sometimes not as easily and cleanly as the most common cold-formed alloy 1018, similar in composition to A36 but with slightly lower carbon content. A36 is generally less expensive than 1018, maybe because of the higher volume of production and distribution.

1018 steel is available in some hardware stores in sheets from 1/16" thick and up, and also in square, rectangular, and round sections up to 1/2". For much better choices of thickness and section, go to a metals stockist. There may be one within driving distance, but if not, any steel you need can easily be purchased online—the downside of that is shipping cost; see Section 4-8.

Before buying anything, consider a free-machining alloy such as 12L14 (trade name Ledloy) or 1215 as a lead-free alternative. The difference in machinability and surface finish compared with 1018, and practically every other alloy, is startling. It can be used in most applications as a direct substitute for 1018. Additionally, 12L14 and 1215 can be case-hardened for some degree of wear resistance.

The only downside with free-machining alloys is shapes and sizes: They are usually available only in bar form (not sheets) in square, round, and hexagonal sections. Thin sheets and rectangular sections can occasionally be found, but don't bank on it.

Table 4-2, at the end of this chapter, shows steel and stainless-steel compositions.

4-15 THINK *HEXAGONAL*

I keep a selection of hex sizes of steel on hand, mostly for turning custom screws, because I don't want to spend the time machining the heads. Hexagon bars work beautifully with the 3-jaw chuck, and in one respect are 100% better than round stock—they are *absolutely resistant to slippage* in the chuck, very helpful when you are trying to cut a coarse thread with a die button.

4-16 SPECIAL JIGS, FIXTURES, AND GADGETS

Don't hesitate to take the extra time to make *interim parts* for work holding and gauging that you subsequently scrap. It's almost always quicker, more accurate, and less frustrating than simply hoping for the best.

4-17 INTERIM PART: EXAMPLE #1

How do you hold a threaded part in the 3-jaw chuck? If the thread is a standard size, run three nuts onto it, two at one end, one at the other. Lock the two nuts together, with the flats aligned, then use the third nut to stabilize the assembly in a 3-jaw chuck. If the thread is a size you don't have nuts for, make a thin-wall plastic or aluminum tube, drilled for a close fit on the thread, long enough to run true in the chuck. Saw-cut the tube along its length (Figure 4-11).

It won't run perfectly, but usually close enough. A less time-consuming way to do this is to use three scraps of aluminum sheet, one under each jaw. (This can also work when holding the OD of a gear.) Whatever method you use, a 4 jaw instead of a 3 jaw can be adjusted for near-perfect concentricity.

FIGURE 4-11 Holding a threaded part. *Left:* Aluminum split sleeve to fit threaded screw to be machined. *Right:* Alternative split sleeve (Delrin).

4-18 INTERIM PART: EXAMPLE #2

Suppose you have an internally threaded accessory that fits a threaded spindle, such as the lathe's headstock spindle, and now you need another accessory with the same internal thread. You can measure the spindle's thread pitch and nominal diameter, but those numbers are only a starting point. If you machine the new accessory to the nominal figures, it might fit loosely, or not at all. However, there's a simple solution: You know that the existing accessory fits nicely, so it can be used as a *go/no-go gauge* for an *interim part*—an externally threaded, portable stub, made from scrap material (probably aluminum) that mimics the target spindle *exactly* (Figure 4-12).

Now test-fit the threaded accessory on the aluminum go/no-go gauge while it's still in the lathe chuck. If it's too tight—good—increase the thread depth a thousandth or two each pass until it fits perfectly, at which point you know that the stub exactly matches the target spindle. Now you can machine the new accessory for a precise fit on the stub. This sounds like a lot of effort. So it is, but it's less effort than machining by dead reckoning, getting it slightly wrong, and having to start over.

FIGURE 4-12 Matching a threaded part. *Left:* Original internally threaded accessory. *Middle:* Interim (aluminum) interim go/no-go part, externally threaded. *Right:* Finished internally threaded product in steel; the arrow points to the central hub, a separate component that was added after the threads were machined.

4-19 IS IT OK TO USE A FILE ON LATHE WORK?

Yes, it certainly is. Filing is the easiest way to "break" sharp edges on shoulders and to taper the end of a turned part in preparation for threading with a die. You can also use a file (or a pad of emery paper, with drop cloths to protect the lathe) to smooth surface imperfections, and to correct minor variations in diameter over the length of a turned part—even to remove the last few tenths of a mil to give a perfect fit of a shaft in a bearing. This can sometimes be more predictable than taking a shaving cut with the knife tool. To use a file on lathe work, don't just lay it on the rotating material—it needs to be stroked gently back and forth in the usual way. "Gently" is the word. "Concentrate" is the other important word: You will be working close to the chuck's flailing jaws, which can propel a file toward you at bullet speed—so use files with handles, not bare tangs.

4-20 WHAT TYPES OF FILE?

The short answer is not the ones from the local hardware store, unless they happen to stock Swiss pattern files. Standard off-the-shelf files are often good for general shop use, but the mill bastard cut is too coarse for precision work. You might get by with an American second-cut or (better) smooth-cut file, but those are not always available. For machine work, I use Grobet files, all #2 cut. (Swiss files are classified by cut number, from 00 to

6, coarsest to finest.) My favorite is the so-called equaling file, 6" long with parallel edges, one of which is "safe," no teeth.

4-21 CLEAN AS YOU GO

Chips and machining residue can have a disastrous effect on the accuracy and finish of a part you are machining. In the shop, we are typically aiming for dimensions accurate within a mil—one thousandth of an inch, 2 thousandths at most. That's equivalent to a metal particle so small you can hardly see it—but imagine what it could do to concentricity if it's trapped between workpiece and chuck. Commercial shops use a high-pressure air line for routine cleaning (but they take care not to blow chips into sliding parts such as chuck jaws and machine ways). Air lines can spread a lot of oil mist and debris around, not desirable in most small shops. A better way is to use disposable paintbrushes. Wash them in hot soapy water.

4-22 REMOVING BROKEN TAPS

This is next to impossible for most of us. Tap extractors usually break before enough torque can be applied to remove the broken item. Commercial shops use carbide ball mills or EDM (electrical discharge machining), but that's usually not possible in the small shop. It's usually a case of starting from scratch on a fresh workpiece, unless—lucky day—you can leave the broken tap in place, then drill and tap in an alternative location nearby.

4-23 CUTTING OFF

It was mentioned in Chapter 2 that cutting off will cause more problems than almost all other lathe operations combined. That's maybe a slight exaggeration, but cutting off does come with a degree of unpredictability. On one day, it goes beautifully; on another day, exactly the same cutting tool—on the same material, lubricated in the same way—squeals, digs in, cuts at an angle, etc. So much so, you might have to reach for the hacksaw to finish the cut, hoping there's no one watching. (Seriously, if you have to do that, protect the lathe bed with scrap wood.)

All of the above comes into sharp focus it you are trying to make a nice-looking custom washer: One side looks great, because you finished it with a knife tool; the other looks to have been gnawed rather than cut—and it's not even flat. See Section 4-26 for any easy way to fix that problem.

4-24 A CUTTING-OFF CHECKLIST

What you need to bear in mind:

- Make the cut as close to the chuck as possible.
- Set the cutting edge on, or *very slightly* below, the centerline.
- Set the cutting tool square to the centerline.
- *Lock the saddle* before making the cut.
- Start with a *low spindle speed*, about 1/4 of what you used for turning the part; think 100 rpm, max., at least initially.
- Check everything for tightness and minimum overhang, "stick-out" not more than two to three times the diameter.
- Don't power-feed the cross-slide until you're sure.
- Experiment with and without cutting fluid.

4-25 CARBIDE CUT-OFF TOOLS MAKE IT EASIER

Carbide insert cut-off tools have been around for years at prohibitive prices, but the landscape has recently changed for the small-shop machinist. For around $50 at 2021 prices, you can buy a cut-off blade, a special holder, and a 10-pack of inserts. I have tried this type of tool on two or three different steel alloys, and the results were better than I was ever able to achieve using the standard Series 100 cut-off toolholder with an HSS blade. Figure 4-13 shows the carbide insert tool in action, maybe the first step in making a steel washer (it could have been, if I had remembered to drill a hole before taking the shot).

FIGURE 4-13   Carbide insert tool slicing off an 0.05" thick steel disc. Close-up of the insert is at right.

4-26 MAKING A BETTER WASHER

Even with the carbide insert cut-off tool shown in Figure 4-13, there's no guarantee that its inside surface will be flat enough to serve as a washer. Its surface finish will also be questionable. The good news is that both problems can be easily corrected.

Make a mandrel by facing a scrap of any material (e.g., PVC) large enough to provide support (Figure 4-14). If you need to trim the perimeter of the washer, center it by leaving a nub of material as shown in the figure. There's no need for the nub if all you want to do is flatten the washer surface.

Apply double-sided adhesive tape to the support surface; then press the (degreased) washer firmly in place (Figure 4-15). The material used in the photo was 3M molding tape, ordinarily used to attach trim to automobiles. It is thicker and more compressible than ideal, but it works fine. There are tapes designed specifically to hold workpieces in applications like this, but they are sold only in large, very expensive, rolls. In practice, any double-sided tape will do, even carpet tape. All it has to do in this instance is withstand a very light facing cut. Remove the finished product by gentle prying. If that doesn't work, try heating it with a hair dryer (heat softens most adhesives of this sort).

FIGURE 4-14   Scrap PVC rod with locating nub.

FIGURE 4-15   Washer attached with double-sided tape. Inset: Washer produced by the process described.

You may be wondering whether a washer made this way will be uniformly thick. In my experience, it will be within 1 or 2 thousandths, even with the squishy molding tape shown in Figure 4-15.

Keep double-sided tape in mind for other light finishing operations, especially milling, where the workpiece is too thin and flexible for normal clamping, and also in cases where the entire surface has to be exposed for machining.

In cases where precise thickness of a washer or spacer was called for, I have used liquid adhesives instead of double-sided tape. One adhesive that has worked well for me is cyanoacrylate super glue. (Contact cement can also be used, in a pinch, but it takes minutes versus seconds to firm up.) The super glue film is very thin—practically zero—so thickness can be measured from washer surface to mandrel face using a depth gauge. Another way to measure thickness is shown in Figure 4-16. Here, the mandrel is smaller than the spacer OD, so calipers can be used in the ordinary way.

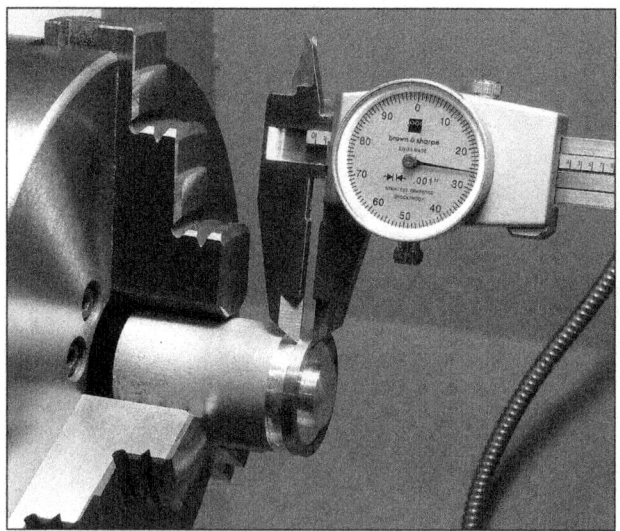

FIGURE 4-16 Measuring thickness of the workpiece. The washer shown here is an unfinished steel spacer on a scrap aluminum mandrel, smaller in diameter to allow measurement using calipers. The ragged inner rim resulting from cutting-off will be removed by light facing cuts.

To separate the workpiece from the mandrel, simply tap it with a soft dowel (wood, brass or plastic). The glue bond fractures easily, but will usually have adequate shear strength for light facing operations.

4-27 MAKING VERY THIN WASHERS

Parting off from a rod is good only for washers up to, say, 0.05" thick (and then only if your cut-off blade is sharp and true). *So what do you do when the project calls for shim washers thinner than 0.05"?* One answer, if you don't mind the delay, is to order them from a supplier such as McMaster Carr. There's the cost factor, too—with tax and shipping, they will be $2 apiece or more.

Instead, use shim stock, something every shop needs from time to time. The usual choices are plastic (vinyl or polyester of some sort), brass, 1018 mild steel, and stainless steel (Figure 4-17). The suggested method works for all material types. Sets of various thicknesses ranging from 0.0005" and up are available from a few online suppliers. It's worth shopping around for small sheets if you can find them, say, 4" x 6" or 5" x 5". Expect to pay about $10 for a set of plastic shims, a lot more for other materials. If you can't find sets at reasonable prices, buy one sheet of 0.004" or 0.005" and double up as necessary.

FIGURE 4-17 Shim washers call for a special process. Here are brass 0.005" and stainless-steel 0.01" washers made in just a few minutes using the most basic equipment.

Cut the shim into little squares, comfortably larger than the OD of the washer. Accuracy is not important. With thin materials, you may find it more convenient to fold the material concertina fashion, leaving it uncut (Figure 4-18). This sometimes doesn't work so well with stainless shim, which is often too brittle to withstand sharp bends. Use hand shears, even scissors, instead.

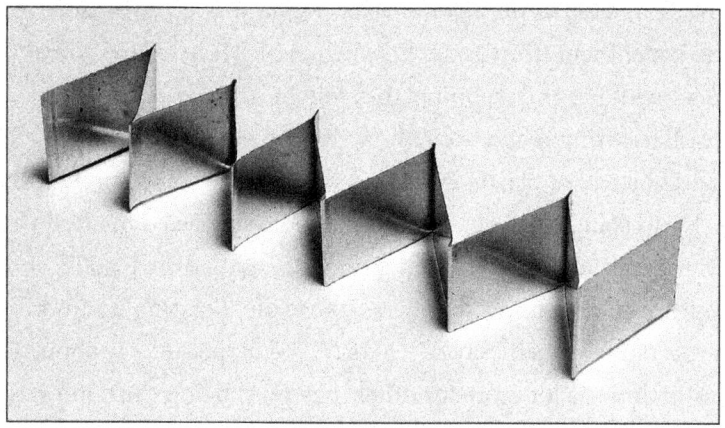

FIGURE 4-18   A Z-folded stack can be easier to handle. This avoids the necessity of handling a pile of very thin metal squares. Use this method for soft, thin materials—some plastics, brass, annealed steel, anything that can be overbent.

Using any scrap material on hand, construct a drill jig on the lines of Figure 4-19. If you reuse the jig, "repair" it by inserting a scrap of sheet metal between the bottom plate and the washer stack. Use material thick enough to hold the stack firmly together, allowing the drill to pierce the lowest shim cleanly. Insert the washer stack, taking care to center it under the hole in the top plate; then drill through.

FIGURE 4-19   A simple drilling jig. This was made from 1/4" aluminum scrap.

For finish-machining, there are two choices, A and B, both calling for a short length of bar stock from the scrap pile. The ID of the washer is the only factor in deciding whether to machine a nub (A) on the end face, with internal threading (Figure 4-20), or (B) an externally threaded stub (Figure 4-21).

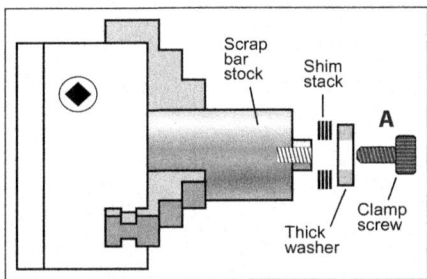

FIGURE 4-20 Washer spindle: Choice A. This is suitable for large ID washers.

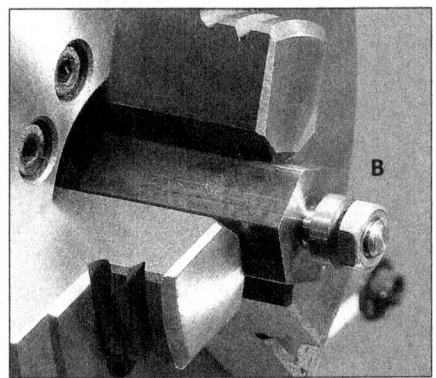

FIGURE 4-21 Washer spindle: Choice B. This is suitable for any washer ID.

In both cases, the smooth portion of the neck *exactly matches* the inside diameter of the washer stack. This means that the internally threaded nub (A) should not be used if its diameter is too small to be threaded for a sensibly sized clamp screw, say, #8.

(You might be tempted to forget the refinement of a smooth neck to locate the washers, simply having them rest on the clamp screw threads. *Not a good idea!* Thin shim stock will slide off-center into the thread "valleys.")

The key factor that applies to both Figure 4-20 and Figure 4-21 is a *thick washer* of the same ID as the nub or stub. In Figure 4-21 (B), the washer is 0.15" thick, resting on an unthreaded neck only 0.125" wide. This guarantees that even a single washer, no matter how thin, will be pressed firmly against the end face of the chucked bar.

The final step is very straightforward—clamp the stack firmly, and then turn to size.

4-28 KNURLING OVERVIEW

Knurling is a hit-or-miss process for even the most experienced. It is complicated enough to get a whole chapter to itself, Chapter 8. As a passing mention, if a project has any knurling, no matter how insignificant it seems, *do it up front*, before any other operation if you can. Get it out of the way before doing anything else! That way, you won't have to redo the entire part if the knurling goes wrong.

4-29 4-JAW VERSUS 3-JAW CHUCKS

Most lathes sold today come with at least a 4-jaw "independent" chuck, one reason being that this type of chuck is less costly to manufacture than the self-centering 3-jaw chuck (which gets more thorough treatment in Chapter 7). The good thing about the 4-jaw chuck is that its jaws are indeed independent, meaning that it can hold very short or odd-shaped items just as easily as round bar stock. (It doesn't do well, however, on hexagonal shapes, for which a 3 jaw is the obvious choice.) The other good thing about the 4 jaw is that it can be adjusted to hold practically anything with great accuracy—meaning, for example, that a drilled hole in an odd-shaped workpiece can be aligned precisely on the lathe axis.

4-30 WHAT IS TIR?

TIR, meaning total indicator reading, is the difference between max and min readings on a dial indicator held against the rotating workpiece. Other phrases with similar meaning are "total indicator runout" and "full indicator movement." Round stock—truly round—can be adjusted in a 4 jaw for practically zero runout, in other words, less than 0.0005" TIR. This is better than you can expect with a typical self-centering chuck, which may have much greater runout. Most do better than, say, 0.005" TIR, but note that with most 3 jaws, the TIR varies with *different workpiece diameters*.

4-31 WHAT SIZE OF 4-JAW CHUCK?

Bigger is not always better. An 8"-diameter chuck, for example, can weigh almost 30 lbs, which is not easy to handle. In addition, its jaws will be about 1" thick, not well suited to typical model shop work. Once in a while, an 8" chuck is necessary, but for everyday use you might want a friendlier size. I went to the other extreme—a 4" chuck, light enough to be installed and removed with one hand, shown in Figure 4-22. The chuck here is mounted on a D1-4 camlock backplate, purchased separately. (The most difficult part of the project was finding a source for the backplate: Machining it to fit the chuck took a matter of minutes.)

FIGURE 4-22   Small 4-jaw chuck.

4-32 CENTERING AN ODD-SHAPED WORKPIECE

First, you need to have some feature on the workpiece that you can define as its center, to be set in line with the lathe axis. This might be a fully drilled hole, a center-drilled hole, or simply a pencil mark on the face. Centering a drilled hole is straightforward if you have a length of close-fitting drill rod or dowel (Figure 4-23).

With the setup of Figure 4-23, you snug each jaw gently on the workpiece, then rotate the spindle by hand to detect and correct any drag on the dowel. Finally, fully tighten the four jaws. Instead of the dowel, you could use a standard "dead" center in the tailstock. Even better is a live center, which rotates along with the workpiece (Figure 4-24). Sometimes, this can help by bringing into view a hidden, now easily detectable, misalignment at the point of contact.

FIGURE 4-23   Using a dowel in the tailstock chuck for centering.

FIGURE 4-24   Live center aiming at a center-punched mark. Here the jaws have been reversed for greater capacity.

One thing to bear in mind when working with a 4 jaw is that the threads are coarse, meaning that the slightest movement of the T-handle displaces the workpiece by an unexpectedly large amount. In the chuck shown in Figure 4-22, the threads are about 6-1/2 TPI, and so a full turn of the T-handle moves the jaw more than 1/8". (The larger the chuck, the coarser the threads: on my 8" chuck one turn of the handle moves the jaw 0.2".)

4-33 CENTERING A ROUND WORKPIECE

First, before putting the workpiece in the chuck, see if it *really is round*. You might be surprised: Use calipers or (better) a micrometer to check the diameter at various points around the circumference: With ground stock, there should be little variation, but on unfinished bar stock you may see ± 0.002" or more. This means you could waste a lot of time adjusting the chuck for zero TIR—it simply isn't possible.

Set the workpiece in the 4 jaw, centering it as best you can by eye. Snug the jaws; then turn the spindle by hand to check for concentricity with a dial indicator (Figure 4-25). (I do this often enough to have a dedicated QCTP holder for my indicator, preset at the height of the lathe axis.)

FIGURE 4-25 Testing for concentricity with a dial indicator.

Here's one way to do the centering:

1. Adjust the jaws (snug only) to bring the runout to less than full scale on the indicator; then note the high and low readings.
2. Rotate the indicator bezel to set the *scale zero* midway between high and low.
3. Rotate the chuck to align the indicator plunger with one of the jaw-pairs, horizontally opposed at front and back (it doesn't matter which pair you choose).
4. Adjust the horizontally aligned jaws to bring the indicator needle to zero.
5. Rotate the chuck 90° to align the second pair of jaws with the plunger.
6. Adjust the second pair of jaws to zero the indicator needle.

This should bring the workpiece close to zero TIR. Check by hand-rotating the chuck another revolution or two. Make minor adjustments as necessary.

Now is a good time to remember how coarse the jaw threads are. When you are chasing that last thousandth of TIR, you should not be thinking of turning the T-handle so much as backing it out imperceptibly, just *taking the pressure off*, followed by a similarly tiny pressure application from the opposing screw.

Finish the job by fully tightening the jaws, in pairs; then check again for TIR.

4-34 COPYING MORSE AND JACOBS TAPERS

Once in a while, we need to taper a spindle to match exactly a tapered socket on an accessory such as a drill chuck. This is a task most machinists would prefer to leave on the back burner, ideally forever, once they've looked at the numbers and found that the taper angle for a Morse taper #2 shank is 1° 25' 49". There is no way to achieve that sort of precision with the compound scale—the best you can do is estimate to the nearest quarter of a degree. To be "self-holding," to be capable of transmitting torque, your shop-machined taper doesn't have to be a perfect fit in the target socket—but it does have to be close.

The way to achieve that is to *copy an existing taper* using a dial indicator as shown in Figure 4-26. The sample taper should be *precision-ground* with good surface finish and have a *parallel section* long enough to be held reliably in a chuck.

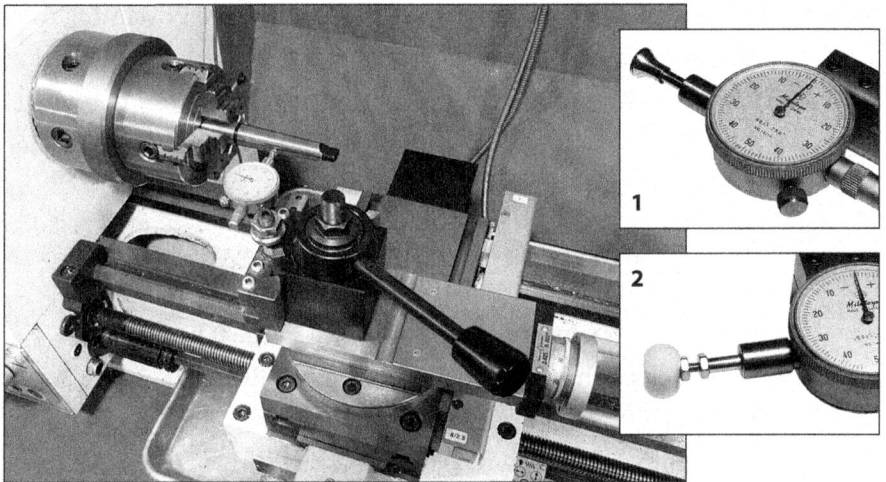

FIGURE 4-26 Setting the compound angle to machine an MT2 taper. Photo 1 shows a flat-bottomed indicator shoe, which may work better in this application than the standard round nose. A better possibility, shown in photo 2, might be a shop-made Delrin follower, also shown in Figure 4-27.

If you cannot find a sample taper with a parallel section for the chuck, there are three alternatives:

1. **(Morse tapers only)** Start with a blank machinable arbor with only the taper portion hardened. Machine the unhardened portion parallel using an *adapter sleeve*, outer to fit the lathe's spindle taper, inner for the machinable arbor, e.g., 5MT to 2MT. The taper in Figure 4-26 was machined that way.

2. Find a sample taper that is center-drilled at both ends (most are, but you need to ask). Install the taper between centers.

3. If you have a spare tailstock center that is center-drilled at the Morse taper end, make a seat for the 60° point at the "business end" by center-drilling a scrap of metal in the chuck.

Very important for taper copying is a smooth-acting compound, with the *gib tightened* to eliminate side-to-side wobble. For good surface finish, choose a free-machining alloy such as 12L14. Also be sure the cutting tool is in good shape. Advance the compound using a balanced grip with your fingers on the handwheel—cranking it with the handle causes wobbling.

For a Morse taper sample held in the 4 jaw, as shown in Figure 4-26, here is a suggested procedure:

- Leave enough of the parallel section exposed to allow centering, as described in Section 4-33. Aim for zero TIR.
- Check concentricity at the outboard end of the taper.
- If the indicator showed zero TIR on the parallel, and more than, say, 0.002" at the outboard end, the chuck is not "pointing straight." Adjust and repeat until satisfied that the error is acceptably small, *or:*
- Set the sample taper between lathe centers (which you will have checked for true alignment—yes?).
- When satisfied with the alignment of the sample, set the compound angle to about 1-1/2° (good as a starting point for all Morse tapers). Leave the swivel screws no more than snug.
- If the dial indicator has a standard ball-end tip, make sure that the tip is at the lathe center height plus or minus a few thousandths.
- Replace the ball-end tip with a flat-bottomed shoe, if you have one available, Figure 4-26, photo 1 (a flat surface at the point of contact helps reduce concerns about indicator height).

- Even better than the flat shoe is a shop-made vertical rod made of Delrin or other easy-sliding material, Figure 4-26 photo 2, also Figure 4-27.

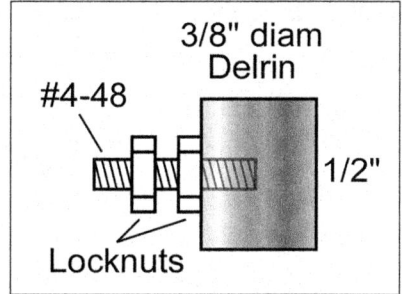

FIGURE 4-27 This type of follower slides easily and is not sensitive to height.

- Note that most dial indicator shafts are threaded #4-48. You can probably do without a special tap, because a #4-48 screw can usually be run into a #4-40 hole if the material is plastic. All you need is a long #4-48 screw and a couple of lock nuts.

- With the follower touching the sample taper, run the compound from one end to the other while watching how the needle moves.

- The aim is zero movement of the needle, highly unlikely at the first attempt. Adjust the compound angle in tiny amounts by gentle tapping with a dead-blow mallet. Check the needle deflection each time.

- After as many iterations as it takes to achieve close to zero deflection, tighten the swivel screws in preparation for machining your copy taper.

- If you are not satisfied with the surface finish on the copy taper, consider very light filing using a fine file (see Section 4-20).

- Mark the copy taper all over with a fiber-tip pen; then push it into the target socket. Rotate the seated taper; then remove for inspection. High spots, if any, will show as bald areas.

- Rechuck the taper to remove high spots by filing (probably no need for supercritical attention to TIR at this stage).

4-35 MEASURING HOLE SIZE

If you have just drilled a hole, you know approximately what its diameter is—the nominal diameter of the drill plus a few thousandths, depending on drill sharpness, chuck alignment, and hardness of the workpiece. A *reamed* hole has a more definite ID—exactly the nominal diameter of the reamer, plus maybe two or three 10-thousandths.

Unknown holes are a different matter. You can measure using the inside jaws of a caliper, but that's only good to within ± 0.002" or so. For more definite information, you need a hole gauge of some sort. *Be wary of using drill shanks* to check inside diameters! The shanks are usually smaller than the nominal size of the drill. A better way is to machine a precise-fitting plug gauge from scrap material and then check its OD with a micrometer (Section 4-36). Two types of commercially available hole gauges are *small hole* (Figure 4-28) and *telescoping* (Figure 4-29).

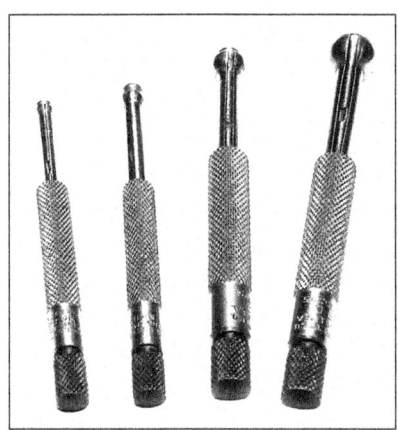

FIGURE 4-28 Small-hole gauges from 1/8" to 1/2" ID.

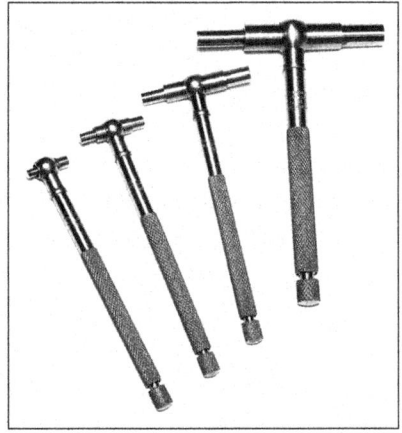

FIGURE 4-29 Telescoping hole gauges from 5/16" to 2-1/8" ID.

A small-hole gauge typically has a hardened split ball, the OD of which is expanded by a taper that is pulled into it by screw action.

The smallest of those shown in Figure 4-28 can be sized to fit any hole diameter from 0.125" to 0.2", the largest from 0.4" to 0.5". To use this type of gauge, place it in the hole, expand it to a snug fit, remove it, and then measure its diameter with calipers or (better) a micrometer. Repeat this two or three times to be sure of a reliable measurement.

The smallest of the telescoping gauges in Figure 4-29 covers the ID range from 5/16" to 1/2", the largest from 1-1/4" to 2-1/8". Care is needed to achieve reliable numbers with the telescoping gauge. Here's how:

- Release the rods by unscrewing the knurled cap at the lower end of the handle.
- With thumb and forefinger, compress the rods into the central casing; then hold them there by retightening the knurled cap.
- Insert the gauge into the hole with the handle on the hole's centerline (or close to it).
- Release the rods; then wiggle the handle back and forth a few degrees to be sure the rods are properly in contact with the bore.
- Retighten the knurled cap; then remove the gauge.
- Measure across the rods with calipers or a micrometer.

It is a good idea to repeat the process a few times—you might be surprised by the spread of readings.

4-36 MAKE A PLUG GAUGE FOR BETTER PREDICTABILITY

Commercial shops usually have sets of precisely ground "pin gauges" from about 10 thousandths diameter up to an inch or so. Most small shops rely instead on the gauges shown in Figures 4-28 and 4-29, but there are times when greater accuracy is a must.

The answer is to make your own *stepped plug gauge* using nothing more than a well-honed knife tool and a 0-1" micrometer, Figure 4-30.

Starting with a scrap of steel about 1-1/2" long, turn 3/4" or so *exactly* to the desired hole diameter. Next, turn the outermost 3/8" down to minus

0.002", and then further reduce the first 3/16" of that length to 0.005"undersize. (These numbers are arbitrary and will vary depending on the target hole diameter.)

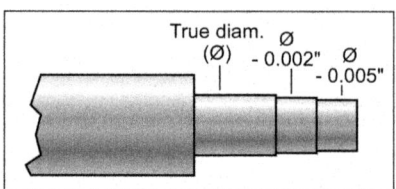

FIGURE 4-30   Shop-made stepped plug gauge.

The great thing about the multistep gauge is that it gives *advance notice* that you are nearing the right size, as compared with a single-diameter plug that (Oh, no!) goes into the bore and wiggles more than you'd like. If the first step, only, of your plug gauge enters the hole, you can safely open up by 0.002", and thereafter proceed by half-thousandths increments on the cross-slide.

4-37 COUNTERSINKING HOLES

For more on the special-purpose drill-point countersink, see Section 2-28. General-purpose countersinks are available with various included angles, the most common being 82° for US hardware and 90° for metric (Figure 4-31). A 120° countersink can be very useful for removing the unwanted center rim from a face-turned disc (Section 4-40).

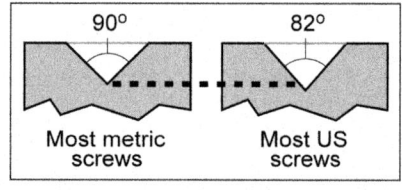

FIGURE 4-31   The two common countersink angles.

Using a countersink is not as obvious as it looks!

In Figure 4-32, bits 1, 2, and 3 are multi-flute bits, which have a tendency to chatter if run too fast or with insufficient pressure. Making decisive contact with the workpiece hole can sometimes help, and so also can cutting fluid when working on steel. Bits 4 and 5, said to be "zero flute," tend to run more smoothly, with very little chatter. For most users, zero flute is the first choice every time.

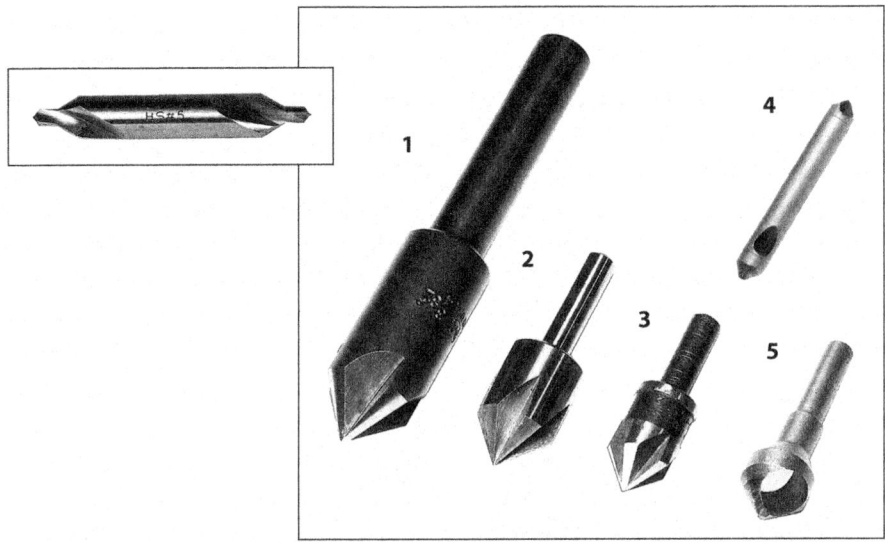

FIGURE 4-32 Countersink bits. The drill-point 60° countersink (*above left*) is used to drill an accurately located pilot hole and also to prepare the ends of a workpiece to be turned between centers. The other general-purpose countersinks here have an included angle of 82° to suit most US flat head screws.

4-38 THERE'S NO SUCH THING AS A SIMPLE JOB

There is a process—a specific order of events—to even the most basic lathe job, such as making a threaded stud. First, you reduce the workpiece diameter, and then turn a shoulder, bevel the cut end, and cut the thread. Using a die, or cutting with a single-point tool? Can you grip the stock well enough to stop it from backing into the chuck under a heavy drilling or cutting load? Or from slipping when die threading? Even worse if it's a double-ended part—the first part might be easy, but how do you chuck the finished end to work on the other? Threads are easily damaged, so you need to plan for it.

Check all measurements and calculations; then recheck them (*measure twice, cut once*). Don't just glance at your part drawing, even if it's a sketch you've just done. Take another careful look. Try to visualize the entire process; don't paint yourself into a corner.

Don't start the motor until you're sure everything is tight and solid—chuck/faceplate, tailstock chuck, tailstock taper, cross-slide and compound gibs, QCTP clamping, toolholder screws. Make sure the workpiece is firmly gripped. Finally, don't unchuck the workpiece until you're sure it's done.

4-39 PLANNING: EXAMPLE #1

Figure 4-33 is a stub axle calling for both ends and the flange to be concentric within ± 0.0005".

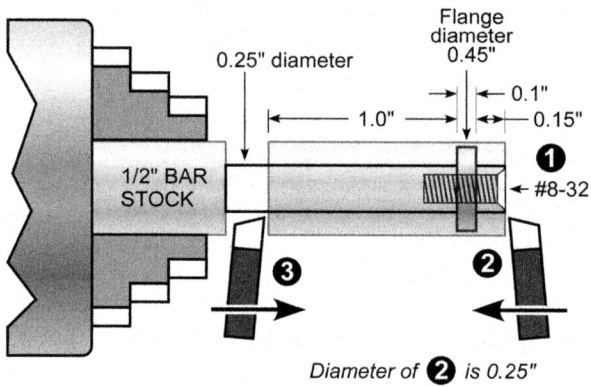

FIGURE 4-33   Stub axle with flange.

If you have a first-rate toolroom lathe with reliable collet chucks, you might choose a different route, but here is an order of events that will work with any lathe:

1. Drill, tap, and countersink at point 1 in Figure 4-33.
2. Make a surface-finishing cut on the end face at point 1 with a right-hand knife tool tilted a few degrees counterclockwise.
3. Cut the outer 0.15"-long x 0.25"-diameter stub down to the flange, point 2.
4. Finish the flange's outer surface.

5. With a parting-off tool, cut a groove at point 3 wide enough for a left-hand knife tool.

6. With the knife tool tilted a few degrees clockwise, cut the 1"-long x 0.25"-diameter portion down to the flange; then finish the flange's inner surface.

7. Part off at point 3.

The above sequence takes care of all the drilling, tapping, and countersinking operations with the rough bar stock held firmly in the 3 jaw. If instead you did those operations after the piece has been parted off, holding the 1" portion in a chuck, your nicely finished spindle could easily be damaged by the chuck jaws.

What if the parted-off surface at point 3 is rougher than the job calls for?

The answer: Make a plastic or aluminum sleeve, 0.5" OD x 0.25" ID, slit as described in Section 4-17, and then finish the spindle end with a knife tool.

4-40 PLANNING: EXAMPLE #2

This example describes how to face-turn both sides of a 1-3/8"-diameter x 3/8"-thick disc with a 1/4"-diameter center hole. The disc could be a gear blank or simply an ornamental part. If it is the starting point for a pulley, the Vee would be roughed out before cutting off.

With 1-1/2"-diameter bar stock in the 3 jaw, face-turn the outer end surface. Drill the 1/4"-diameter center hole (if this has to be exactly 1/4" to fit a spindle, drill undersize; then ream). Cut the disc from the bar stock, slightly overthick, either by parting off on the lathe or by sawing.

The disc will be slightly thicker than the desired 3/8", unevenly thick, and its inner surface will be rough, even if cut off on the lathe.

How can the disc be made true-running, with the specified thickness and with a good surface finish all around? This can be done by making a temporary "sacrificial" hub (see part 1 in Figure 4-34), from scrap bar stock, with a 1/4"-diameter nub, a little less than 3/8" long. Drill and tap, say, #8 or M4.

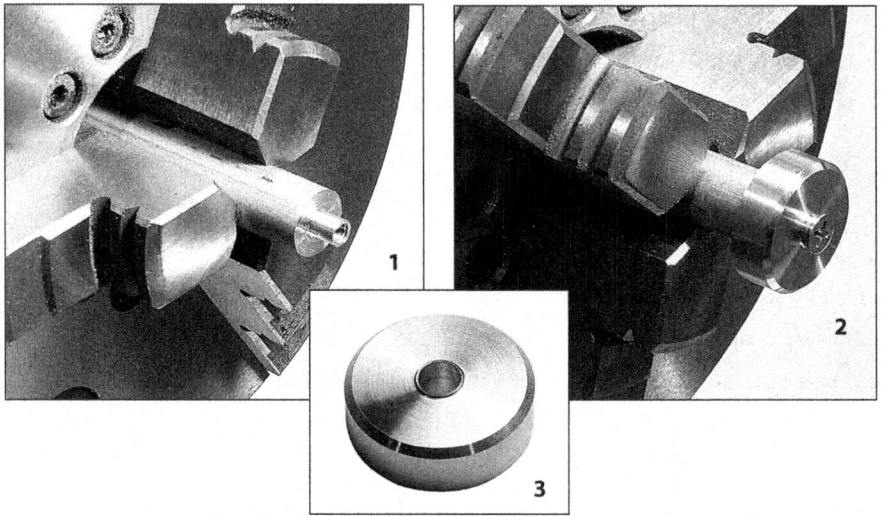

FIGURE 4-34 Part 1 is machined from scrap material, tapped #8-32 in this example.

With a flat head screw, attach the disk to the hub, rough face out (2). Skim the underside of the flat head screw to get as close as possible to the hub with the knife tool. Turn the disk OD to 1-3/8"; then skim the outer surface for the desired disc thickness, leaving only a thin rim at the hub (3), but don't go in so tight that the rim breaks down. Also, do not risk this procedure on a large, heavy disc. *Safety glasses are a must.* Remove the surplus metal on a *drill press*, with a drill larger than the upstanding rim, or use a hand-held countersink bit (ideally 120°).

There's one more part to "job planning": If you are making a number of similar discs with the same size of center hole, even if they might have different outer diameters, *rough-cut them all* to size before making the temporary hub. The hub is a one-time setup. You can machine it with practically zero TIR, but you *cannot reinstall* it in a 3 jaw with the same concentricity. This is because bar stock usually isn't truly round, and the 3-jaw chuck may have its own repeatability issues.

There are alternative ways of doing the job. For instance, Figure 4-35 shows a live center pressing into the center hole, providing enough traction for turning if the hub face is coated with adhesive or double-sided tape, as described in Section 4-26.

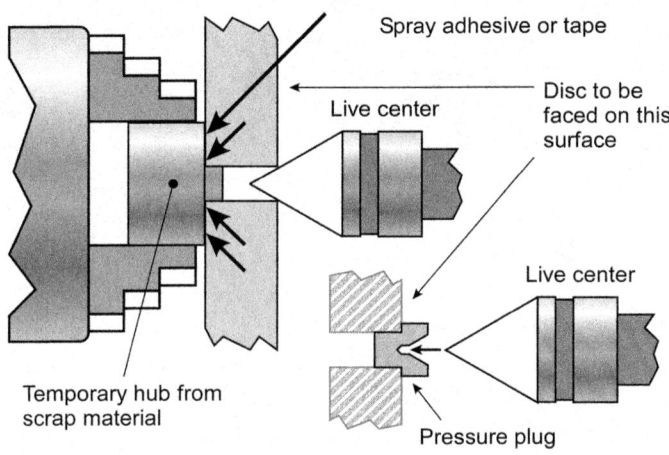

FIGURE 4-35   Using a live center to press the disc against the hub. Setup (1) may be adequate for light work on small discs. A pressure plug (2) keeps the knife tool well clear of the live center and is also a visible indication of the facing cut's inner limit.

If the disc was cut from accurately dimensioned (and nicely finished) sheet material, such as aluminum jig plate, there may be no need for any face cutting, in which case the disc can be strongly attached to the temporary hub with a screw, as shown in Figure 4-36. Alternatively, the hub can be made with a threaded extension, as shown earlier in Figure 4-21.

4-41 PLANNING: EXAMPLE #3

This is one of many ways to make a heavy disc, rough-cut from 5"-diameter aluminum bar (Figure 4-37).

The spec called for ± 0.001" TIR and ± 0.002" axial wobble at the rim.

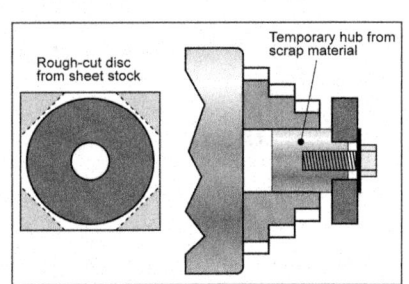

FIGURE 4-36   Attaching a disc that does not need to be face-cut.

FIGURE 4-37   A 5"-diameter, 1/2"-thick disc attached to a stub axle; the photo at the upper left shows three tapped holes for attachment of the axle.

In the following procedure, the rim and one of the sides (fairly light-duty jobs) are finished with the disc attached to a *prefinished* stub axle:

1. Make a stub axle, as shown in Figure 4-38, with the outer section about 1/4" longer than the disc thickness—we use this extra length to check TIR (step 7 below).
2. Before parting off the axle, use a tool-post drill (Chapter 9, Figure 9-17) to drill three clearance holes at 120° in the axle flange. Note the *exact location of the cross-slide*, for reuse in step 5.
3. Set the rough-cut disc in a 3 jaw with external jaws; then face-cut the outer side.
4. Drill, then ream the center hole for a snug fit on the stub axle.
5. With the tool-post drill at the same cross-slide setting as in step 2, drill and tap three holes at 120° for the attachment screws.
6. Attach the finished face of the disc against the flange; then secure it with three screws.

7. Set the disc/axle assembly in a 4 jaw; then adjust for minimum TIR by indicating the protruding end of the axle (Figure 4-38).

8. Face-cut the outer side and rim.

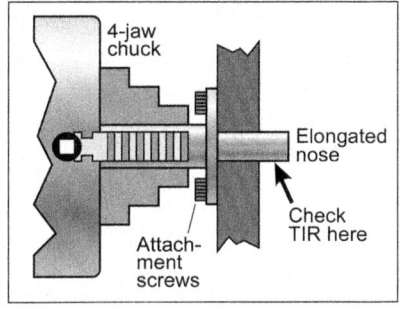

FIGURE 4-38 Stub axle.

4-42 A LOW-COST MODELING METHOD

A trial run can avoid mistakes! Like everything else in engineering, the finished product is the best compromise of competing factors.

A reliable way to be sure you have the dimensions right is to model the entire project in materials that are easy to work—cardboard, wood, or (much better) an "engineering material" like rigid PVC (Figure 4-39). This is a low-cost plastic, available in rod, bar, and sheet from many suppliers, including McMaster Carr.

FIGURE 4-39 Machining PVC is quick and easy.

PVC machines beautifully at high spindle speeds, with no significant cutter wear—but watch out for overheating. Blanks of the material can be rough-cut to size in seconds using a table saw with a wood-cutting blade.

Aside from solid-modeling CAD and 3D printing, this is the fastest way to a realistic fit, form, and function model (with a lot less learning time, too). Another bonus you get from working in PVC is the ability to practice machining procedures with the least possible risk. *One caveat:* When heated, PVC expands four times more than steel—and it cools more slowly—so take care when checking measurements. Reliable screw cutting can also be an issue, especially with fine-pitch threads.

For experimental work, I keep a stash of 3/4"-thick dark-gray PVC (Type 1), plus a few oddments of bar stock. Where extra thickness or composite shapes are needed, PVC can be bonded in a minute or two using plumbing adhesive.

4-43 SHORT WORKPIECES AND CHUCKS

Suppose you have drilled a round bar in the chuck, then parted off a short length of it, say, 1/2" or so, to make a collar. The outer face of the collar will likely be perfect, but not so the parted-off face. *Why not reverse the collar in the chuck, then skim it with the same knife tool you used on the outer face?* Sounds easy—until you try it—then see that the collar won't run true without some form of parallel spacer between it and the chuck face. You could try pushing the collar up against a *temporary* spacer of square bar stock, then tighten the chuck. If you happen to have the right size of square material on hand that's certainly a possibility, with one obvious problem: "temporary" means you need to remove it before machining the workpiece, so there is nothing to resist cutting forces. *Result:* The workpiece is pushed back, and/or will wobble some more. Much better would be custom spacers for your various chucks, 3 jaw and/or 4 jaw, as in Figure 4-40.

I have an assortment of these in various thicknesses, from 1/2" to 1/16". They were shop-made from aluminum or PVC sheet (McMaster Carr). For a 3-jaw chuck, drill the sheet material with holes at 120°, diameter larger

FIGURE 4-40 Shop-made spacers from sheet material.

than the jaw thickness, then bandsaw three slots. For a quicker way to make several spacers with the one setup, turn 2" or 3" disks of the material, then use a rotary table and an end mill to cut the slots. (For the 6" chuck shown, the end mill diameter was 3/4".)

4-44 SCALES AND MICROMETERS

In the woodshop and around the home, you probably use scales graduated in fractions of an inch: 1/4", 1/8", 1/16", etc. The two factors here are *Familiarity* and the *Instructions* you're working to. The same scales are not as useful in the metal shop, where measurements are more likely to be in thousandths of an inch (mils) or millimeters. This calls for different measuring tools, starting with the "engineer's scale." The most useful choice is generally a flexible 1/2"-wide scale, 6" or 12" long with decimal/metric graduations: 1/10" and 1/50" one side, 1 mm and 1/2 mm on the other (Figure 4-41). These are between 0.015" and 0.02" thick, which gets you closer to the workpiece than the three times–thicker rigid scale. Decimal/metric scales like this are worth shopping for—usually online only.

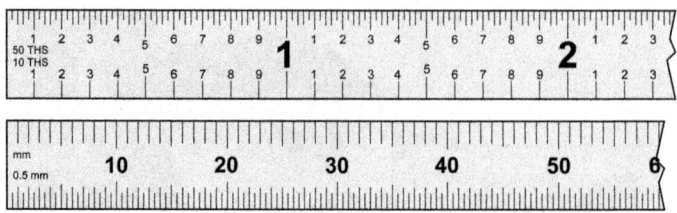

FIGURE 4-41 Typical inch/metric engineer's scale, front and back.

With a "fiftieths-of-an-inch" scale like that shown in Figure 4-41, you can directly measure to the nearest 0.02". With interpolation, you can estimate to ± 0.01", 10 thousandths, but that's the practical limit. On the metric scale, we are limited to the stated resolution of the finest graduations, namely 0.5 mm. To do better than that, we need a scale with a more precise *interpolator*, in other words, a vernier scale. This was invented almost 400 years ago by a French mathematician, Pierre Vernier. It has been in worldwide use ever since.

How a vernier scale works is an interesting mystery to many of us, including experienced machinists. Imagine a generic "decimal reading" caliper graduated in any unit you wish—feet, inches, or centimeters.

In Figure 4-42, the jaws are closed, with both zeros aligned. Note that 10 on the vernier scale spans 9 on the main scale, a one-count difference that is typical of all vernier scales.

Now separate the jaws so that the vernier 0 is about three-quarters of the way between 0 and 1 on the main scale.

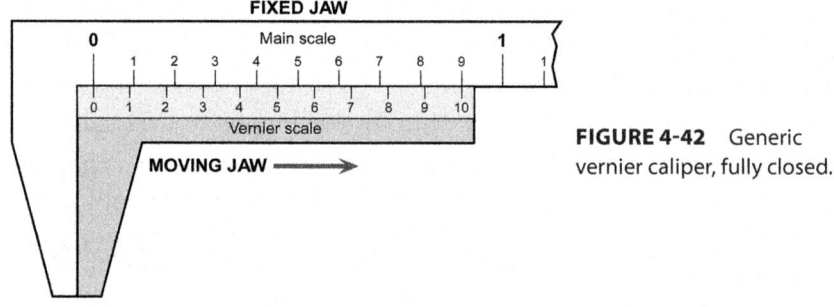

FIGURE 4-42 Generic vernier caliper, fully closed.

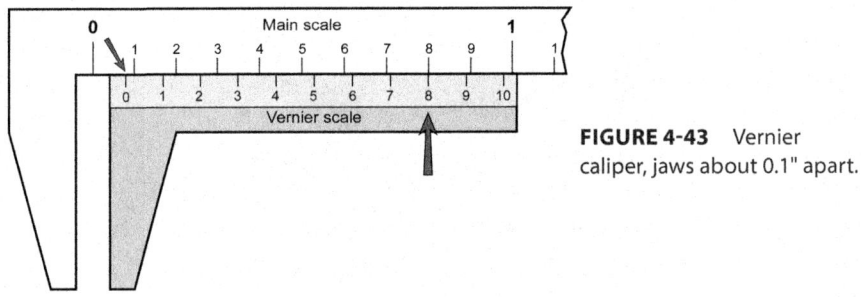

FIGURE 4-43   Vernier caliper, jaws about 0.1" apart.

The question now is, *How can we know more exactly where the vernier 0 is, which tells us how far the jaws are apart?* The answer is to look along the vernier scale for a line that coincides with a line on the main scale. In Figure 4-43, it's line 8, so the jaws are separated by precisely 0.08 on the main scale. In practice, the scale would be graduated in decimal fractions of an inch or in centimeters, so the 8 could be read as a specific measure of distance.

Another question that crops up when you try to use an unfamiliar vernier scale is, *What is its resolution?* Sometimes, this is announced (for instance, 1/1000"), sometimes not. If not, determine the unit of measure (U) represented by the smallest division on the main scale, usually 0.025" or 0.05". Then count the number of divisions (V) on the vernier scale. Finally, divide U by V.

In the real-life vernier of Figure 4-44, the upper scale is in inches, divided into tenths, with further subdivisions (U) of 0.025". The vernier scale has 25 divisions (V), so the resolution U/V is 0.025"/25 = 0.001", as marked (arrow). In the photo, Example 1, the vernier 0 is just to the right of 2 on the main scale, meaning a little greater than 0.2". Further to the right, there is perfect coincidence of the two scales at vernier 20, so the jaws are separated by 0.2" + (20 x 0.001") = 0.220".

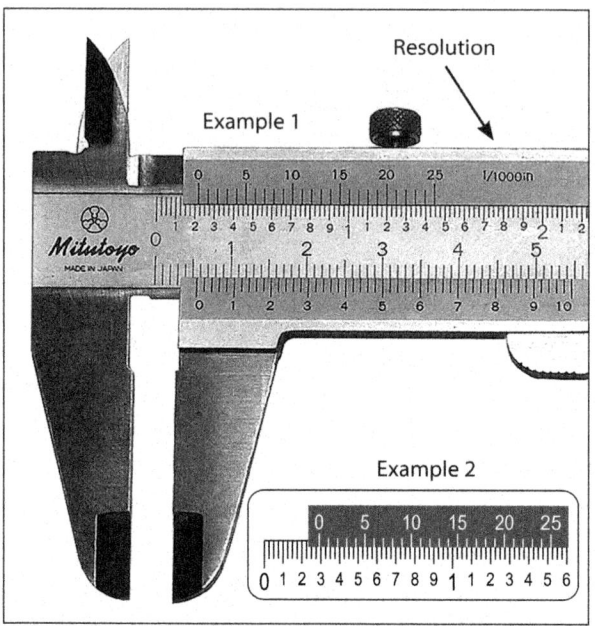

FIGURE 4-44 Inch/metric vernier caliper.

In Example 2, Figure 4-44 inset, the vernier 0 is to the right of the *third sub-division* between 0.2 and 0.3, and there is coincidence at vernier 17, so the total measurement is 0.2 + (3 x 0.025) + (17 x 0.001) = 0.292".

Back to the photo in Figure 4-44, the lower scale (metric) is quite different. For one thing, its resolution is not announced, so we need to figure it for ourselves. With the jaws closed (not shown), 50 divisions (V) on the vernier span 49 divisions on the main scale—the usual one-count difference. Each division of the main scale represents 1 mm (U), so the resolution U/V is 1 mm/50 = 0.02 mm (a shade finer than 0.001"). In the photo, the vernier 0 is about midway between 5 and 6 mm on the main scale, meaning that it lies in the range 5.5 through 5.7 mm. Looking to the right, there is perfect coincidence at vernier 6, so the correct value is 5.6 mm. In other words, numbers on the vernier scale *directly report* tenths of a millimeter, with subdivisions of 0.02 mm (for instance, if the coincidence were to be at the next vernier line to the right of 6, the value would be 5.62 mm).

The takeaway here is that *step 1* in all vernier measurements is "visual interpolation" to determine roughly where the vernier 0 lies, followed by *step 2*, careful checking for coincident lines.

Both steps call for a steady hand, good lighting, and good eyesight, all factors that spawned easier-to-use technologies—first the dial caliper, then the direct-reading digital scale.

So why trouble with a centuries-old measurement method? Three answers: First, because there are great deals out there for vernier calipers and height gauges. Second, there is nothing to go wrong with a vernier caliper—it's coolant-proof and has no battery and no fragile, easily jammed rack and pinion. Third, as will be described in a moment, because almost all *standard inch micrometers* use a vernier scale to increase resolution from thousandths of an inch to 10-thousandths, 0.001" to 0.0001".

In the inch-reading micrometer in Figure 4-45, the numerals 0, 1, etc., above the left-to-right datum line represent intervals of 0.1", with four subdivisions of 0.025" (these are not numbered—your first hint that the micrometer calls for visual interpolation, just like the vernier caliper). The rotating sleeve is numbered **0** 1 2 3 4 **5** 6 7 8 9 **10** 11 12 13 14 **15** 16 17 18 19 **20** 21 22 23 24 **0**, each division representing a 1-mil interval. In the figure, the total distance visible along the datum line is 5 x 0.025" = 0.125".

If the 0 mark on the sleeve happened to coincide with the datum line, the measurement would be *exactly* 0.125". However, the line is between the 2- and 3-mil lines on the sleeve, in fact just a shade beyond 2. If we are working only to the nearest mil, we would say that

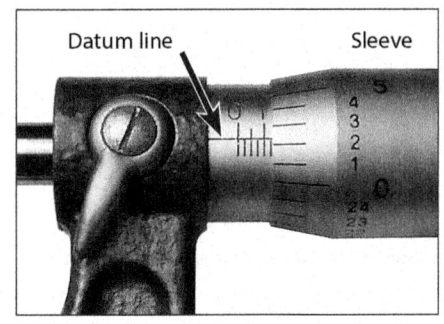

FIGURE 4-45 Inch-reading micrometer.

the measurement is 0.125 + 0.002 = 0.127". If we need to be more precise, we lock the sleeve, then roll the micrometer forward to view the vernier (Figure 4-46). Here, the only vernier line coincident with any line on the sleeve is 3, indicating that the measurement is 0.1273".

4-45 COMMON STEEL ALLOYS

Table 4-2 shows the chemical composition of a variety of common steel and stainless steel alloys.

General purpose steels such as 1018 and A36 are classified as "simple alloys" of iron, carbon, manganese and silicon. Tool steels have in addition varying percentages of chromium, molybdenum, vanadium and tungsten.

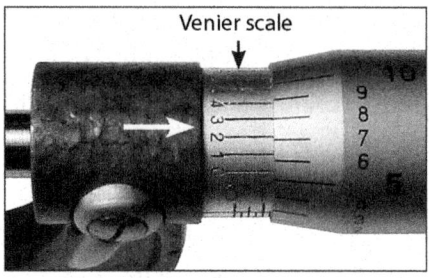

FIGURE 4-46 Micrometer vernier scale. The vernier 3 aligns with 8 on the sleeve, so the value is 3 x 0.0001" greater than the thousandths value measured in the ordinary way. Total: 0.125" + 0.002" + 0.0003" = 0.1273".

Stainless steel comes in three main categories: *austenitic, martensitic,* and *ferritic.* The main difference between all stainless steels and simpler alloys is their high chromium content, plus—in some cases—more exotic elements such as tantalum and niobium.

Austenitic steels are the most common. They are non-magnetic (unless work-hardened). They cannot be hardened by heat treatment. Martensitic steels, on the other hand, are *magnetic,* and can be hardened and tempered much like non-stainless low-alloy steels. Ferritic steels, not listed in Table 4-2, cannot be hardened by heat treatment; however, they are *magnetic* and more ductile.

4-46 CAST IRON

Cast iron, sometimes referred to as "gray iron", is used for machine tool bases, engine blocks, gears, flywheels, stoves, etc.

Its *carbon* content of 2.5% to 4% by weight is more than 10 times that of common steel alloys. This is the main reason for its very different machining characteristics (powder instead of chips). Other typical constituents of gray iron are *silicon* 1% to 3%, *manganese* 0.15% to 1%, *sulfur* 0.25% max, and *phosphorus* 1% max.

TABLE 4-2 Percentage by weight of steel alloy constituents (plus iron, totaling 100%)

STEEL TYPES	AISI ref	Carbon	Manganese	Silicon	Phosphorus	Sulfur	Chromium	Molybdenum	Vanadium	Other
Construction	A36	0.08–0.29%	0.40–1.20%	0.15–0.40%	0.04% Max.	0.05% Max.	—	—	—	—
General purpose	1018	0.13–0.20%	0.30–0.90%	0.15–0.30%	0.04% max.	0.50% Max.	—	—	—	—
	1117	0.14–0.20%	1.00–1.30%	—	0.04% max.	0.08–0.13%	—	—	—	—
Higher strength	1045	0.43–0.50%	0.60–0.90%	0.15–0.30%	0.04% max.	0–0.05%	—	—	—	—
	4140	0.36–0.46%	0.65–1.1%	0.15–0.04%	0.035% max.	0.04% max.	0.75–1.2%	0.10–0.25%	0.05–0.15%	—
Easy machining	12L14	0.15% max.	0.85–1.15%	—	0.04–0.09%	0.26–0.35%	—	—	—	0.15–0.35% Lead
	1215	0.09% max.	0.75–1.05%	—	0.04–0.09%	0.26–0.35%	—	—	—	—
Tool steels	A2	0.95–1.60%	0–1.00%	0–0.60%	0.030% max.	0.030% max.	4.75–5.5%	0.70–1.40%	0.15–1.10%	—
	D2	1.40–1.65%	0.60% max.	0.30–0.60%	0.030% max.	0.030% max.	11.00–13.00%	0.5–1.2%	0.5–1.10%	—
	O1	0.85–1.05%	1.00–1.40%	0–0.50%	0.030% max.	0.030% max.	0.40–0.70%	—	0–0.30%	0.40–0.60% Tungsten
	W1	0.95–1.05%	0.10–0.40%	0.10–0.25%	0.025% max.	0.025% max.	0.15% max.	0.10% max.	0.10% max.	0.15% max. Tungsten

(continued on next page)

STEEL TYPES	AISI ref	Carbon	Manganese	Silicon	Phosphorus	Sulfur	Chromium	Molybdenum	Vanadium	Other
High-speed steels (HSS)	M2	0.78–0.90%	0.15–0.40%	0.20–0.45%	0–0.030%	0–0.030%	3.75–4.5%	4.50–5.50%	1.75–2.20%	5.50–6.75% Tungsten
	M42	1.05–1.15%	0.15–0.40%	0.15–0.65%	*—	*—	3.50–4.25%	9.0%–10.0%	0.95%–1.35%	1.15%–1.85% Tungsten 7.75–8.75% Cobalt

STAINLESS ALLOYS	AISI ref	Carbon	Manganese	Silicon	Phosphorus	Sulfur	Chromium	Molybdenum	Nickel	Other
Austenitic (typically non-magnetic, unless work-hardened)	303	0.15%	2.0%	1.0%	0.2%	0.015%	17–19%	0.60%	8–10%	—
	304	0.08%	2.0%	1.0%	0.045%	0.030%	18–20%	—	8–10.5%	—
Not heat-hardenable	316	0.08%	2.0%	0.75%	0.045%	0.030%	16–18%	2.0–3.0%	10–14%	0.1% Nitrogen
	316L	0.03%	2.0%	0.75%	0.045%	0.030%	16–18%	2.0–3.0%	10–14%	0.1% Nitrogen
Martensitic (magnetic, heat-hardenable)	410	0.15%	1.0%	1.0%	0.040%	0.030%	1.5–13.5%	—	—	—
	440C	0.95–1.20%	1.0%	1.0%	0.040%	0.030%	16–18%	0.75%	—	—
	17-4PH	0.07%	1.0%	1.0%	0.040% *	0.030% *	15.0–17.5%	—	3.0–5.0%	3.0–5.0% Copper 0.15–0.45% Niobium + Tantalum

Nominal values only—expect wide variation between suppliers. Trace amounts may be present even when no value (—) is specified. Phosphorus and sulfur (indicated by *) are regarded as impurities. AISI is the American Iron & Steel Institute.

CHAPTER 5

Cutting Screw Threads on the Lathe

CONTENTS AT A GLANCE

5-1 SCREW THREAD BASICS

In the metric community, a thread is described by its major diameter x pitch, in millimeters. *Example:* M5 coarse pitch is M5 x 0.8. (Incidentally, 0.8-mm pitch is very close to 1/32", so a #10-32 screw will fit an M5 nut—good to know in an emergency.)

In the United States, a screw is described in terms of threads per inch, or TPI, using an arbitrary number (#) to define major diameter: #0 (0.06"), #1 (0.073"), #2 (0.086"), up to #12 (0.216"—does anyone ever actually use this size?). For larger sizes, the fractional major diameters are listed directly (1/4", 5/16", 3/8", and so on). The thread pitch P is the reciprocal of the number of threads per inch, so P = 1 ÷ TPI. The numbers for P and the major diameter are the starting point for cutting a thread on the lathe (Figure 5-1). Technically, threads are measured by pitch diameter, PD, the diameter of an imaginary cylinder that intersects the thread to give equal peaks and valleys.

Most threads used in the United States and Canada are defined by "Unified" standards: UNC = coarse, and UNF = fine. There is also UNEF = extra fine from #12 and larger.

5-2 SCREW CUTTING ON THE LATHE

Tool geometry is important. Figure 5-1 shows an included angle of 60°. To achieve this, there are two choices of cutting tool: The first is to use pre-ground, ready-to-use HSS or carbide bits. The second is to grind your own HSS bits to 60°, ideally using a fixture.

If you are grinding by hand, check the angle with a "fishtail" gauge (Figure 5-2).

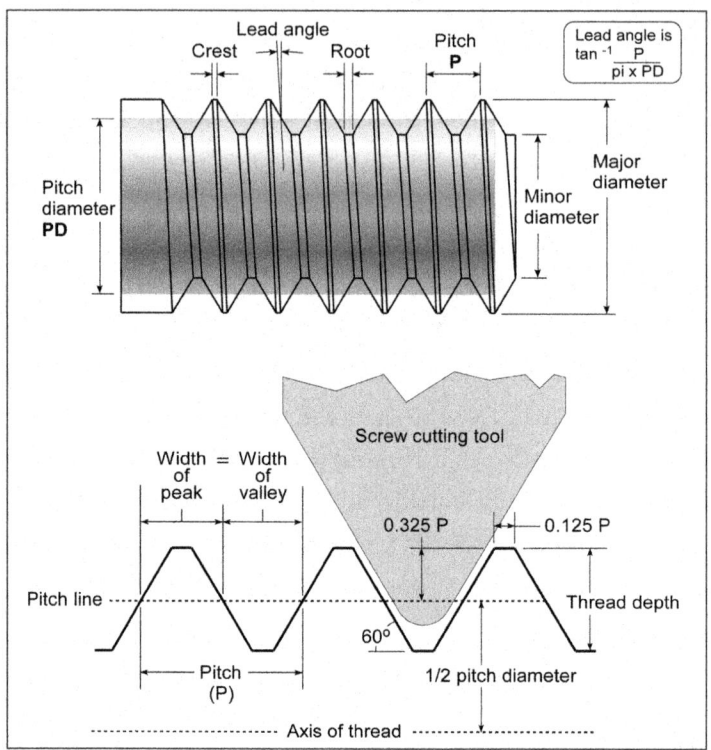

FIGURE 5-1 Unified thread form (UNC, UNF, etc.). Metric threads have the same triangular form but are differently specified. The lead angle, or helix angle, is the angle made by the helix of the thread, measured at the pitch diameter. It is typically between 1° and 4°.

Perhaps surprisingly, you can have flats at the crest and root of the thread, as shown earlier in Figure 5-1, but in real life we are talking about very small numbers—a 20-TPI thread, for instance, can be in spec with a 6 thousandths–wide crest, a more "pointy" thread than you might think from the diagram.

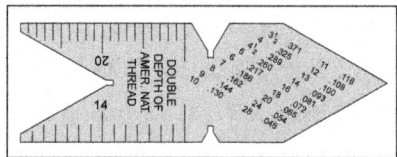

FIGURE 5-2 Fishtail gauge. This is a vintage example, from about 1930, showing the now-obsolete American National thread, superseded from 1950 by the Unified system (UNC, UNF, etc.). It is still useful as a thread profile gauge.

Whatever else, the tool must be ground so that the flanks of the thread are flat from the crest to a "straight-in" depth of about just over half the pitch (actually 0.54P). However, without a microscope we can't know the exact geometry of the tool tip—how flat, rounded, worn? In practice, after grinding the basic 60° shape, most of us just flatten the tip a few thousandths, and that's it.

Pre-ground inserts are usually radiused, so they should not need honing.

5-3 LOW-COST TOOL BLANKS

Round 3/16"-diameter HSS blanks in a square-section shank work well for thread cutting and cost next to nothing (Figure 5-3).

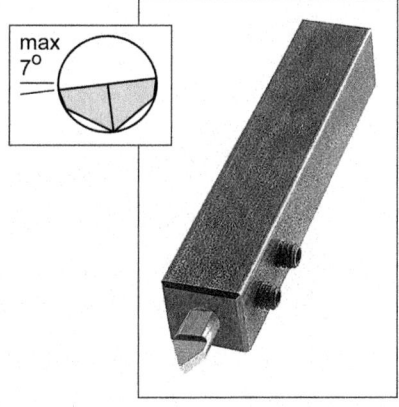

Grind the sides to 30° each side of the centerline, with a flat-top cutting surface. For much better cutting on steel—less tearing and roughness—rotate the bit in the shank for a few degrees of side rake on the left-hand edge (up to 7° won't affect the 60° form appreciably), but you might want to pretilt to 7° before grinding the 60° sides.

FIGURE 5-3   A 3/16" tool bit ground for screw cutting. The inset shows a side rake of 7°.

5-4 THE SCREW CUTTING PROCESS

For screw cutting on the lathe, the saddle is driven along the bed by a lead screw that is coupled to the spindle through a train of external "change gears" (and in some cases, a gearbox) on the headstock. In everyday non–screw cutting operations, an open *split nut* (half nut) in the saddle rides freely along the lead screw. When screw cutting is called for, the split nut is closed onto the lead screw by a lever on the front face of the saddle (Figure 5-4).

The number of threads per inch is determined by the ratio of spindle speed to lead-screw speed. For 20 TPI, for example, the change gears would be set up so that the saddle moves 1" for every 20 revolutions of the spindle.

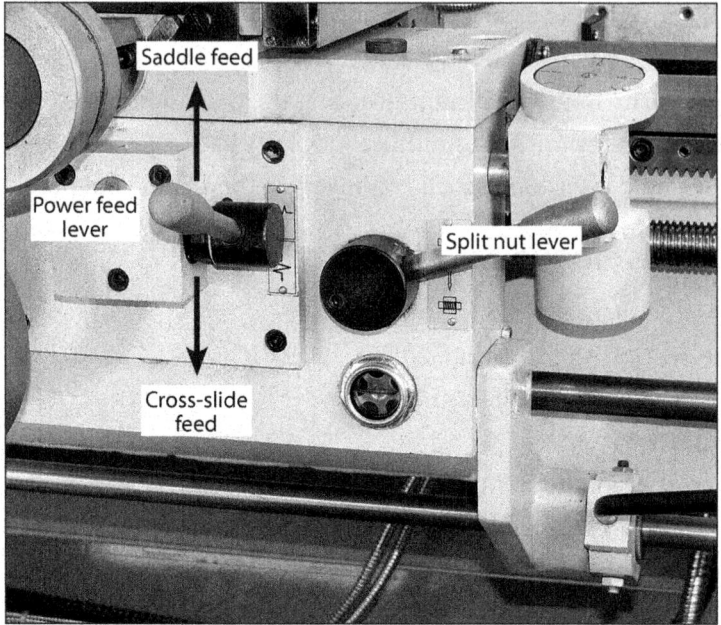

FIGURE 5-4   Typical power-feed and split-nut controls.

The thread is cut in a series of successively deeper overlaying passes, with a depth increment of *only a few thousandths* of an inch each time. *Never* plow into the nominal depth in one go!

The first "cutting pass" is usually only a *scratch cut* to see if the gears are set up correctly. When the first pass is completed, the tool is backed out clear of the workpiece—by retracting the cross-slide—and the spindle is reversed to bring the saddle back to the starting point. The cross-slide is then returned to its former setting, and the tool is advanced a few thousandths by the compound for the next pass. Each successive pass is done in the same way, each with a slightly increased infeed setting of the compound.

In the following notes, we'll assume that the change gears and/or gearbox have been set up according to the lathe's handbook to run the saddle at the appropriate rate for the screw thread you have in mind.

5-5 METRIC VERSUS US THREADS

When you are cutting *metric threads* with a US lead screw (usually 8 threads per inch), the split nut *must remain engaged* throughout the entire multipass process. This means that the saddle is power-driven by the lead screw on both the cutting pass and the return pass.

When cutting *US threads* only, you have two choices:

1. Leave the split nut engaged full-time.
2. Disengage at the end of each cutting pass; then, using the handwheel, return the saddle to the starting point for the next pass.

Many users working on US threads save time by going with choice #2—disengaging the split nut at the end of each cutting pass, reversing the saddle quickly by hand, and then re-engaging, usually by reference to the *threading dial* (Figure 5-5). In non–screw cutting, everyday turning operations with the split nut open, this dial rotates

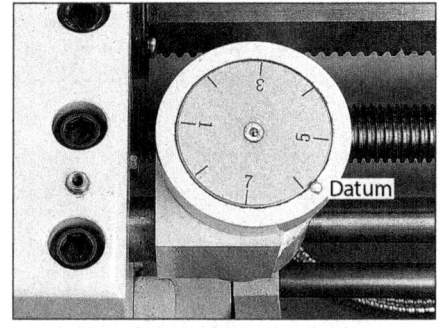

FIGURE 5-5 Typical threading dial.

continuously. When the split nut is closed for screw cutting—on the instant of closing—the dial stops rotating.

For most TPI numbers, every engagement, *including the first*, must be at the point where a *specific line* on the threading dial comes into alignment with the datum mark. If not, the second and subsequent passes will be out of sync. In some cases (see the "visualization" in Figure 5-6), there is a choice of lines for re-engagement, but in every case the process calls for careful timing.

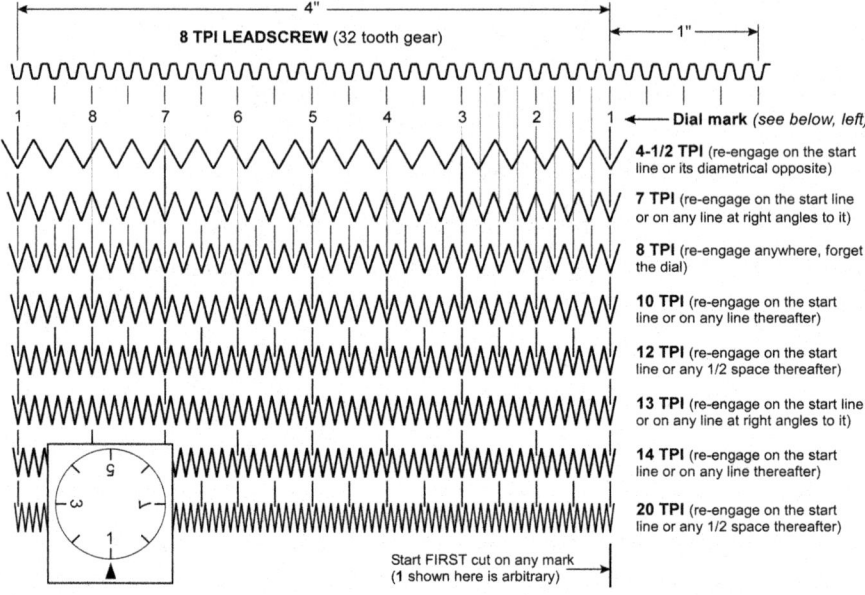

FIGURE 5-6 Re-engaging the split nut using the dial. This visualization shows that with an 8-TPI lead screw, there are more re-engagement opportunities for even-value TPI numbers than there are for odd ones—and fewer opportunities for unusual numbers such as 4-1/2 TPI. This diagram is typical only, and does not apply exactly to any specific lathe.

DISENGAGING THE SPLIT NUT SAVES TIME

If the split nut is engaged full-time, the return trip to the start point will be at the same slower-than-usual speed used to cut the thread. This can be tedious, especially on long thread runs.

5-6 THREAD DEPTH

The cutting process begins with round stock a few thousandths less than the major diameter (see Table 5-1 later in the chapter). At the end of the cutting process, after maybe 10 or more passes of a few mils each time, the thread should have a minor diameter of approximately 0.6 x pitch (the "straight-in" trial depth). For more on this, see Section 5-15.

5-7 THE THREADING DIAL

The threading dial is driven by a pinion that engages the lead screw. (On some lathes the dial assembly pivots outward to disengage when not in use.) The dial is typically secured to the pinion shaft by a central screw. Before any screw cutting operations, the dial should be synced as follows (usually a one-time operation):

1. With the lead screw *stationary* (motor off), engage the split nut. If it does not engage right away, nudge the saddle left or right using the handwheel.
2. Loosen the dial screw; then rotate the dial to align any of the numbers *or* lines with the datum.
3. Tighten the dial screw.
4. Run the motor; then try re-engaging the split nut at any line or number.
5. The re-engagement point may be slightly off due to backlash. If excessive, stop the motor, and re-sync the dial.

5-8 SETTING UP THE COMPOUND (STANDARD METHOD A)

In addition to its need for a 60° pointed cutting tool, screw cutting on the lathe is unlike other turning operations. Although you can, if you wish, feed the tool in at right angles to the work (some do), most machinists prefer to angle the compound so that the tool cuts mostly on a flank (side) of the thread. This can be either the left flank or the right flank, depending on the cutting direction: In the standard setup, "Method A" (Figure 5-7), the saddle motion is right to left, so the tool cuts on the left flank.

The correct angle of the compound relative to the cross-slide is debatable, but many go with 29° because at that angle the cutting tool is just clear of the right flank of the thread, yet close enough to clean up the flank without rubbing.

No matter what the compound angle, the tool post is always angled so that the cutting tool is *exactly at right angles to the lathe axis*. One way to check this is to hold a fishtail gauge (Figure 5-2, shown earlier) against parallel bar stock in the chuck, and then advance the tool into the Vee.

5-9 SETTING UP THE COMPOUND (METHOD B)

Some users, myself included, prefer the setup shown in Figure 5-9. Here the compound is again at 29° to the cross-slide, but in the other direction. In B, the cutting tool is working mostly on the right flank of the thread.

Saddle motion is the key difference between the two setups—toward the left in Figure 5-7, to the right in Figure 5-9.

5-10 WHY B VERSUS A?

If you have ever tried screw cutting with the conventional setup A, you will know that the big question is how to end the cutting pass, because the spin-

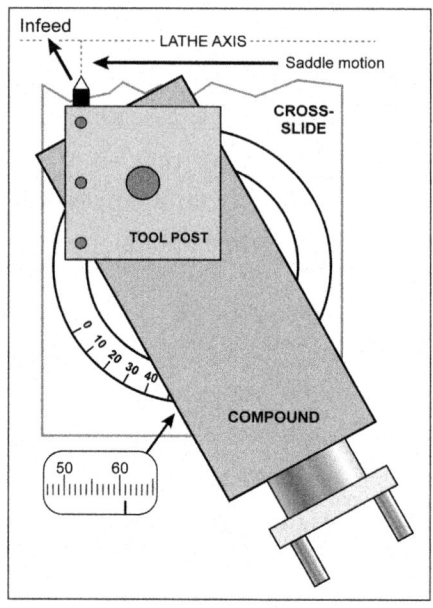

FIGURE 5-7 Compound set for 29° infeed, Method A.

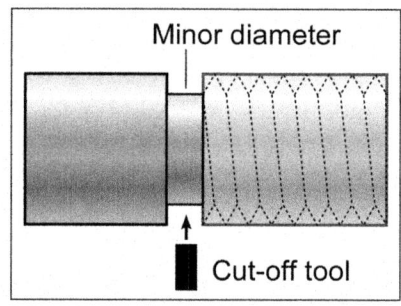

FIGURE 5-8 Runout groove for threads cut using Method A; start groove for Method B.

dle can't be relied on to stop abruptly at exactly the right spot (and if it did, the tool would be deeply into the work as the cutting speed falls to zero—near fatal to carbide inserts). The usual remedy is a runout groove wide enough for wiggle room either side of the screw cutting tool (Figure 5-8).

Sadly, that's only a partial answer. You also need to imagine how the tool arrives at the runout groove. Is it creeping along so slowly you have a good shot at stopping in the groove? If so, that's a much lower speed than is good for the tool or the work.

If instead it is going at a more reasonable speed, you need to stop the spindle in advance, then crank it home by hand-turning the chuck. Another way to end the cutting pass is to let the tool run up to the groove and then retract the cross-slide a full turn at *lightning speed*. You might think this is a shade too heroic, but some machinists do it every day. The problem goes away to some extent if the lathe has a spindle brake, but not many do.

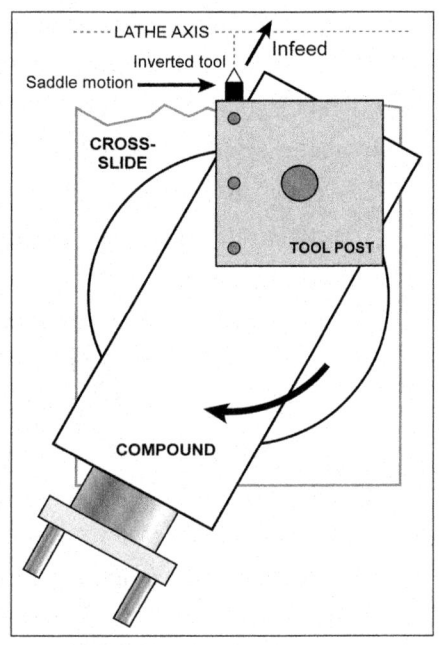

FIGURE 5-9 Preferred 29° setup, Method B.

All of the above has added up to a horror story for machinists over the past 100 years. Consider Method B instead.

5-11 METHOD B IS LESS STRESSFUL

Method B takes care of the runout problem by *inverting the cutting tool and driving the saddle to the right*. This runs the tool toward the tailstock, into open space. It really is that simple, but there are a few points to bear in mind:

- With Method B, tool inverted, instead of a runout groove *you need a start groove,* a gap into which the tool can be advanced in preparation for the next cutting pass (Figure 5-8).

- To drive the saddle to the right, the *lead screw* must be *run in reverse*. This occurs automatically when the spindle is reversed. (But, for *left-hand threads*, the spindle and lead screw go in opposite directions, Section 5-19.)
- Screw-on chucks sometimes do the opposite—unscrew—unless safety clips are installed. (This isn't just theory—I have seen it happen.)
- If the tool bit is angled, as shown earlier in Figure 5-3, it should be rotated in the other direction to put the cutting edge on the right flank of the thread.
- If the angled compound is too close to the chuck jaws, and you can't take care of that by pushing the tool further out of the toolholder, you may have to use an angle less than 29°, or even a *straight-in* feed instead (in which case forget the compound—do everything with the cross-slide only).
- You can set a *saddle stop* to ensure a repeatable start location. However, one thing you cannot do is run the saddle back under power to the stop; this is practically guaranteed to displace the stop, or worse. Instead, for US threads, disengage the split nut when the tool is clear of the work; then reverse the saddle carefully to the stop by handwheel; *or* for metric threads, leave the split nut engaged, reverse the motor—stopping it short—and then hand-rotate the chuck the last couple of turns for a gentle arrival.

The following sections assume Method B.

5-12 MAKING A "SCRATCH CUT"

Cut a start groove as shown in Figure 5-8. To begin the screw cutting process, retract the compound by two or three turns to be sure there is forward motion available. *Don't touch the compound again* until you are setting up for the second pass. Move the saddle by hand to set the cutting tool at the starting point of the thread.

Advance the cross-slide to bring the point of the tool into light contact with the workpiece, just grazing it; then *zero the cross-slide dial* (Figure 5-10) or the DRO X scale. To eliminate backlash error, be sure that contact is made with the cross-slide moving in one direction only—inward. If you think you have applied too much pressure, don't back it out by just a couple of thousandths: Back the handwheel a full turn, and then bring it in again.

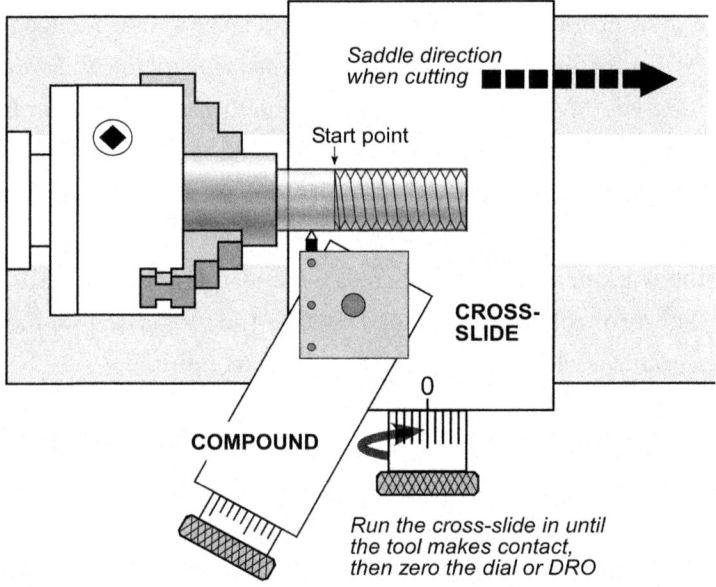

FIGURE 5-10   Setting up for the scratch cut. With the cross-slide zeroed, move the saddle by hand to set the tool at the start point; then (*optional*) set the saddle stop. With this setup the spindle runs backward, clockwise facing the chuck, cutting tool inverted.

The aim in Figure 5-10 is to make a scratch cut that can be checked with a thread gauge, Figure 5-11. (Blacking the workpiece with a fiber-tip pen makes the scratch cut more visible.)

Why only a scratch cut? This is to make sure, before it's too late, that the gear setup is delivering the desired thread pitch. If not, correct the gearing and repeat.

1. Initiate the scratch cut by running the spindle *slowly in reverse* (between 50 and 100 rpm).

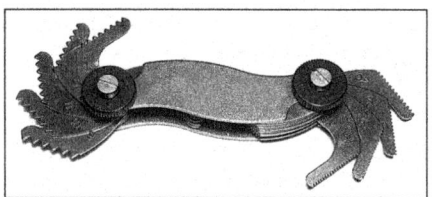

FIGURE 5-11   Thread gauge for Unified (nonmetric) threads, UNC, UNF, etc.

2. With the cutting tool at the start groove, engage the lead screw by closing the split nut (half nut) on the saddle. If you are cutting a US thread, make a note of the dial indication—line or number—when the split nut engaged (when the dial stopped turning). *Depending on the TPI value, it may be necessary to re-engage at exactly the same dial indication (see Figure 5-6, shown earlier).*

3. **Very important:** Immediately on completion of the scratch cut, and *every cutting pass thereafter,* when the tool is clear of the workpiece, retract the cross-slide by *one full turn* of the dial (probably 0.100" on the DRO) so the tool is *clear of the workpiece* when it is returned to the start point.

4. Choose *one* of these two options:

   • **Must-do for metric threads, also usable for US threads.** Stop the spindle, but leave the split nut engaged. Reverse the motor to return the saddle under power, stopping just short of the saddle stop; then complete by hand-cranking the chuck.

   • **For US threads only.** Stop the spindle, then disengage the split nut and return the saddle by handwheel to the saddle stop.

WHAT IF THE CUTTING TOOL BREAKS?

This is a show-stopper, but it doesn't have to be terminal, provided there are two, three, or more cutting passes to go before arriving at the target depth. Recovery is a matter of synchronizing a replacement cutting tool with the existing part-completed thread. Assuming the breakage occurred during a cutting pass, leave the split nut engaged while installing the replacement tool—with its Vee properly aligned and at the correct height. Do *not* touch the workpiece, chuck, or spindle—leave everything in the "cutting direction" throughout to ensure that any lost motion in the split-nut and lead-screw gearing is already taken up. Using a *combination of compound and cross-slide* back-and-forth movements, position the cutting tool for a visually perfect fit in the thread groove. Hand-crank the spindle a few turns in the *cutting direction* to be sure.

5-13 SECOND PASS—SPLIT NUT ENGAGED FULL-TIME

This is the usual procedure for metric threads. It applies not only to the second pass, but to all subsequent passes:

1. Stop the motor after the cutting pass; then retract the cross-slide.
2. Restart the motor in reverse to run the saddle back to the starting point. Stop the motor when the saddle is just to the right of its stop.
3. Move the saddle back to the stop by hand-cranking the spindle.
4. Return the cross-slide to its former position, zero on the dial or DRO. (If you are working without a DRO, watch out for backlash error.)

5. Advance the *compound*, not the cross-slide, by, say, 0.005"*
   (see Section 5-15).

6. Run the motor in the cutting direction to make the second
   pass.

7. Repeat steps 1 through 6. *And if working on steel, don't forget
   the cutting oil.*

*Typical depths of cut per pass vary from an initial 0.005" or so on the compound, progressively down to as little as 0.001", even less. A finishing pass or two with increments of only 0.0005"—or none at all, to deal with the "spring-back effect"—can make all the difference between a too-tight thread and one that runs perfectly.

Table 5-1 shows basic external thread cutting data for common UNC and UNF screw sizes, plus a selection of metric near equivalents. Starting diameters of the uncut material should be a few thousandths under the nominal major diameters listed here.

Trial Infeed Depth is given in two ways: 1. Cross-slide infeed "at 90°", and; 2. The more conventional "at 29°", as measured on the compound dial.

At the "trial infeed depth," check the thread with a nut (or by measuring its pitch diameter); then proceed to the final size. The pitch diameters here are for tight-fitting Class 3 external threads. For standard-fitting Class 2 threads, deduct 0.001"; that's all it takes to make a noticeable difference.

TABLE 5-1   Basic thread-cutting data

Size	Major Diameter (")	Pitch P (")	Pitch Diameter (")	Trial Infeed Depth 0.6 x P at 90° (")	0.69 x P at 29° (")
8-32	0.164	0.0313	0.1437	0.019	0.022
10-24	0.190	0.0417	0.1629	0.025	0.029
10-32	0.190	0.0313	0.1697	0.019	0.022
1/4-20	0.250	0.0500	0.2175	0.030	0.035
1/4-28	0.250	0.0357	0.2268	0.021	0.025
5/16-18	0.313	0.0556	0.2764	0.033	0.038
5/16-24	0.313	0.0417	0.2854	0.025	0.029
3/8-16	0.375	0.0625	0.3344	0.038	0.043
3/8-24	0.375	0.0417	0.3479	0.025	0.029
7/16-14	0.438	0.0714	0.3911	0.043	0.049
7/16-20	0.438	0.0500	0.4050	0.030	0.035
1/2-13	0.500	0.0769	0.4500	0.046	0.053
1/2-20	0.500	0.0500	0.4675	0.030	0.035

Size	Major Diameter (")	Pitch P (") (mm)	(")	Trial Infeed Depth 0.6 x P at 90° (")	0.69 x P at 29° (")
M4 x 0.7	0.157	3.523	0.1387	0.017	0.019
M5 x 0.8	0.197	4.456	0.1754	0.019	0.022
M6 x 1	0.236	5.324	0.2096	0.024	0.027
M8 x 1.25	0.315	7.160	0.2819	0.030	0.034
M10 x 1.5	0.394	8.994	0.3541	0.035	0.041
M12 x 1.75	0.472	10.829	0.4263	0.041	0.048
M14 x 2	0.551	12.663	0.4985	0.047	0.054

5-14 SECOND PASS—SPLIT NUT DISENGAGED

This is the usual procedure for US threads:

1. Following the cutting pass, disengage the split nut, retract the cross-slide, then use the handwheel to back the saddle to its stop.
2. Return the cross-slide to its former position, zero on the dial or DRO. (If you are working without a DRO, watch out for backlash error.)
3. Advance the compound by, say, 0.005" (see Section 5-15).
4. While observing the threading dial, re-engage the split nut at the *exact dial indication* you used when starting the scratch cut (Section 5-12), *or* if permissible, re-engage at an alternate indication (see Figure 5-6, shown earlier).
5. Disengage the split nut when the cutting tool is clear.
6. Repeat steps 1 through 6.

The above procedure applies not only to the second pass, but to all subsequent passes.

BE CAREFUL WHEN RE-ENGAGING!

If you get the re-engagement wrong, it can be a fatal, start-over error, especially if you are several passes into a well-established thread. However, you may not have to wait for a specific number to come up on the dial—for many TPI values, even numbers in particular, there are several choices when timing the re-engagement (Figure 5-6, shown earlier).

5-15 DEPTH OF CUT

Several factors come into play in this:

1. Most run-of-the-mill screws you buy from the hardware store have major diameters well below nominal, in many cases by 0.005", some even more. Expect a 1/4" screw to measure 0.245 ± 0.001". The same applies sometimes even to "quality" screws such as hardened socket head screws from specialty suppliers.

2. As a rule of thumb, for thread cutting on the lathe, start with a major diameter at least 1% less than nominal.

3. In successive passes, feed in the cutting tool a total of 0.69 x thread pitch = 0.69 x P, measured on the compound dial. This is equivalent to 0.6P measured at right angles to the lathe axis: $(0.69 = 0.6/\cos 29°)$.

4. Typical cut depths per pass are cuts #1-2-3 0.005", #4-5 0.004", #6-7 0.003", #8-9-10-11 0.002", #12, and thereafter 0.001" or less. (Thinner stock may need lighter cuts to minimize the spring-back effect.) Overall, you will probably need a few thousandths more than the 0.69P infeed (see Table 5-1), almost certainly not less.

5. As you home in on the final size, take care not to overdo it. A finishing pass or two with increments of only 0.0005" (or none at all) to deal with the spring-back effect can make all the difference between a too-tight nut and one that runs perfectly. See Section 5-21.

5-16 FINISHING THE THREAD

Screw cutting on steel, sometimes other metals, almost always leaves tiny burrs on the outer diameter. These can be removed by a few passes with a fine file with the motor running. If you are looking for a tight fit in a mating part, do several trial fits, filing a little each time.

Also consider using a threading die. Many machinists routinely cut to a depth of 75% or so with the single-point tool, then finish with a die while the workpiece is still in the chuck. This is an excellent way to help the die run squarely on the thread (not so easy if you are threading with a die from scratch).

With large-diameter coarse threads, be aware of the torque necessary to run the die. Not a problem if you are working with hex bar stock, but it takes a lot of clamping force from the chuck to stop round stock turning—so much so there's a danger of damaging the 3 jaw. Take the workpiece to a bench vise instead.

5-17 SINGLE-POINT CUTTING OF INTERNAL THREADS

For those of us who don't machine parts for a living (even for some who do), internal screw cutting can be a nightmare in the same class as knurling (Chapter 8). The main reason for this is very little visibility of the cutting action. This is not too serious if you are threading a through hole—just keep on cutting until you're sure it's through. But now think about threading to the bottom of a blind hole: *How can you be sure of cutting reliably to the full depth without collisions?*

One answer is to use a variant of Method B (Figure 5-12), in which the saddle moves to the right. This shows the other main problem with internal screw cutting—tool overhang. The deeper the hole, the more robust the tool support needs to be. This applies to all internal cutting operations, no matter what the compound setup, Method A or Method B.

The first step in cutting an internal thread is to find a screw that can be used to test-fit the thread you are about to cut. (An equivalent to this in a commercial shop would be a precision GO/NO gauge, Section 5-23.) If the screw size is not one that's commercially available, machine a test screw, then check its dimensions by one of the methods described at the end of this chapter. Secondly, drill or bore a hole a little smaller than the minor diameter of the screw. For standard screws up to 1-1/2" diameter, you can read the approximate minor diameter from the "tap drill chart" you proba-

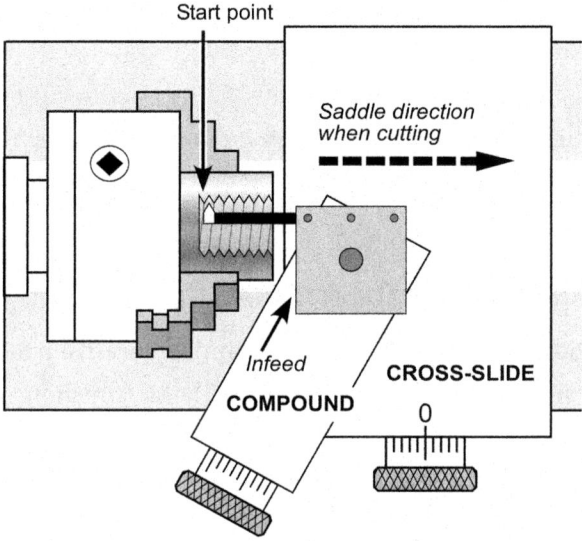

FIGURE 5-12 Internal threading, Method B. In this setup the tool is face-up, and the spindle rotates backward, clockwise viewed facing the chuck. A start groove is always a good idea (see Figure 5-8 shown earlier). The tool cuts on the right flank of the thread.

bly have taped to the workshop wall. Examples, for 5/8-11: hole size 0.531";
3/4-16, hole size 0.688". To be on the safe side, drill a little smaller than the chart reading. If the screw is non-standard, measure its major diameter, D, and pitch, P. Estimate the drill size by deducting 1.2 x P from D. Example, for 1/2-20: D = 0.5", P = 0.05", deduct 0.06". Drill size 0.44".

How small an internal thread is workable? This depends on hole depth, tool size, and rigidity. In most situations, 1/2" is probably the lower limit. For instance, the hole size for 1/2"-13 is just over 0.4", which isn't much space for the cutting tool and its support bar.

In principle, all the preceding notes on Method B apply to internal screw cutting, with these additional thoughts:

- The start point of the thread is almost always in the blind, and you can't just guess where the tool is when the saddle is returned for the next pass. A saddle stop is *highly recommended*, even essential, unless you have a DRO. Even then you might like the reassurance of a physical stop—but don't run into it under power.

- Just as in Section 5-16, you may find it easier to cut to 75% of the thread depth with the single-point tool, and then finish with a bottoming tap—but again, don't ignore the torque requirement.

5-18 FINISHING THE THREAD (INTERNAL)

Remove the burrs with Scotch-Brite or a similar abrasive pad, cleaning up carefully to remove dust particles from the lathe (finishing with a tap is better; see above).

5-19 LEFT-HAND SCREW CUTTING

After glancing through the above, this might seem too hideous to contemplate, but it may not be so bad once you note that in Method B screw-cutting the saddle *always moves to the right.* So the question comes down to this: *How can we cut a left-hand thread with that same rightward motion?*

Answer: Reverse the connection between spindle and leadscrew.

Reversing is possible on most lathes, but not all. If your lathe is not so equipped, it may be possible to insert an idler gear into the external gear train (Section 1-23).

When reverse is engaged, the saddle moves to the right when the spindle rotates forward, counterclockwise as viewed facing the chuck. (This is the exact opposite of what happens to the saddle in non–screw cutting operations.) Figure 5-13 shows the two left-hand configurations.

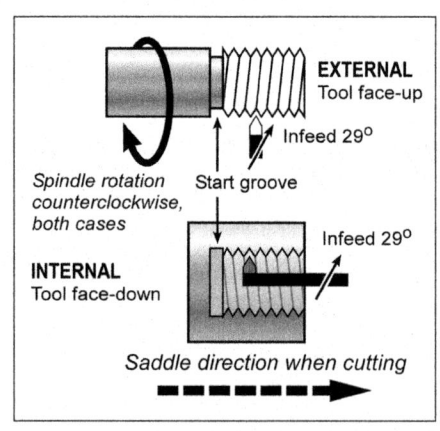

FIGURE 5-13 Left-hand threads.

5-20 HOW SMALL A THREAD CAN I CUT ON THE LATHE?

This is up to you, but keep in mind that small-diameter workpieces flex due to cutting pressure. This can cause the thread depth to vary from one end to the other—a nut that's tight in the first two or three turns may run easier toward the chuck. (I have seen this effect even on 2"-long screws as beefy as 5/16 and M8.) One fix is to use a live center and/or a traveling steady rest. No matter what setup you are using, it is a good idea to minimize the spring-back effect on thin stock by taking lighter-than-usual cuts throughout, making one or two final passes with no change in tool depth. Realistically, anything smaller than #8 or M4 is going to be troublesome unless it's a very short threaded nose on a larger-diameter rod.

For #10 and anything smaller, I usually cut in one pass with an adjustable round die in a tailstock holder; for sizes larger than #10, I use a single-point tool and (sometimes) finish with a die. Even then, threads coarser than 20 TPI can take a lot of effort—sometimes the cutting forces may cause stock slippage even in a fully tightened chuck (just imagine what might be happening to the scroll and the jaws when you really bear down on the T-handle). An excellent fix for this problem, especially with steel, is to use hex-section stock instead of round; see the "*Think Machinability*" box.

"THINK MACHINABILITY"

There is a world of difference in the machinability of 12L14 and 1215, compared with that of off-the-shelf alloys such as 1018. The main downside of 12L14 and 1215 is the limited selection of cross sections—round, square, and hexagon. *Hexagon-section stock*, mostly 12L14 (a leaded alloy), is an obvious choice for making headed bolts, but it's also much better for any high-torque jobs, such as die threading, because it doesn't slip in the chuck and therefore doesn't need such heavy-handed tightening.

Another factor that can noticeably affect fine-pitch screw threads is the sudden change in lead-screw load when the carriage handle goes over top dead center. This is rarely mentioned, but it's real. The fix for it is simply to remove the handle temporarily.

5-21 CHECKING THE THREAD: EVERYDAY METHODS

When the thread is nearing its final size—or so we hope—the usual check is to test-fit a nut on the workpiece. If it's very tight, shave off another 0.0005" or so at a time until the fit is just a hair on the tight side. At that point, a pass with no increase in depth (a "spring cut") should do the job. The nut might be a common or garden variety item from the hardware store or, if we need more precision, a special nut made on the lathe with a thread tap, ideally H1—the smallest of the H limits. (This is a whole topic in itself; see Section 5-27.) An external thread that fits an H1-tapped test hole without excessive force should be usable in any situation.

5-22 CHECKING THE THREAD: MORE EXACT METHODS

Once in a while we may need to cut to a specific pitch diameter, PD, this being the size of an imaginary cylinder intersecting the thread to give equal peaks and valleys (Figure 5-1, shown earlier). Because the cylinder is *imaginary*—a good word for it—PD cannot be measured directly except with special tools that hide the underlying math. The same goes for the minor diameter—not so imaginary, but difficult to measure without an optical comparator.

5-23 COMMERCIAL THREAD GAUGES

Commercial shops usually check external threads with a ring gauge (Figure 5-14). These come in pairs, GO and NO-GO. They sell for about $200 the pair, which puts them over the top for the small shop. They are similar in appearance to adjustable threading dies, the main difference being that they are adjusted in manufacture and then sealed with wax or other tamper-evident material.

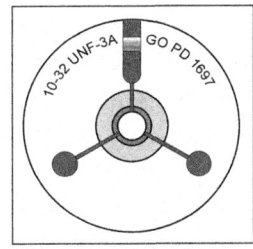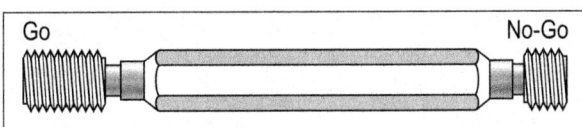

FIGURE 5-14 Commercial thread gauges.

The example shown in the figure is a GO ring gauge for #10-32 set at PD 0.1697"; its NO-GO counterpart (not shown) would be set at a PD of (for instance) 0.1674". Nuts and other internal threads are checked using plug gauges, which are precision-ground screws typically supplied as "two-in-ones" of GO and NO-GO sizes.

5-24 MEASURING PITCH DIAMETER WITH THREAD TRIANGLES

One way of measuring PD less expensively uses thread triangles (Figure 5-15). These are precision-ground wedges, sold as matched pairs for $30 and up. They often come with flexible rubber holders that fit over the spindle and anvil of a standard micrometer (this makes the measurement less of a balancing act than with loose wedges). Set them on either side of any thread from 4 to 80 TPI, measure across the flats with a micrometer, and then read the pitch diameter from the manufacturer's chart—a don't-lose item, not interchangeable with others.

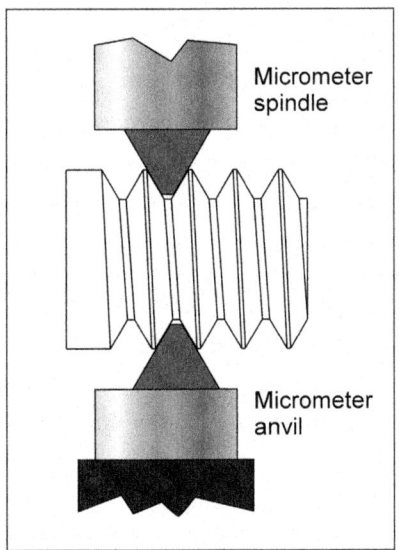

FIGURE 5-15 Thread measuring triangles.

A word of caution: For an accurate reading of PD with triangles (or wires, Section 5-25), the thread form must be a near-perfect 60°, which means using either a carbide insert or precision-ground HSS. Hand grinding without a jig doesn't do it—if the tool angle is fractionally too sharp, the triangles will contact the thread nearer the major diameter than they should and vice versa. (The same requirements for an exact thread profile apply to the screw thread micrometer—see Section 5-26.)

5-25 MEASURING PITCH DIAMETER WITH THREE WIRES

I haven't used my thread triangles for some time now, because I can't find the chart that came with them. The manufacturer is no longer in operation, so they sit in the drawer along with other historical artifacts. What I use now—or did until I bought a screw thread micrometer—is the three-wire method. This uses three wires of identical diameter (within less than ± 0.0001") chosen so they sit proud of the major diameter (Figure 5-16).

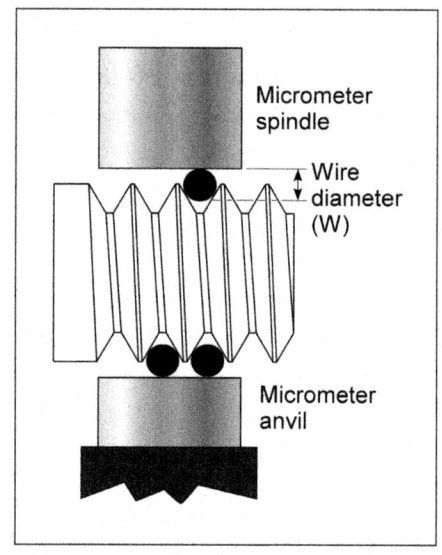

FIGURE 5-16   Three-wire measurement of pitch diameter.

The wire diameter can be anything between 0.56 and 0.90 x P, the thread pitch. A typical commercial wire set comprises 16 sets of three wires—basically 3"-long drill rods—for threads between 3 and 48 TPI. To make a measurement, you place two of the wires on one side of the thread, clamping them with the micrometer anvil; then you insert the third wire into an opposing Vee (this sounds easy, but you will wish for three hands). Measure across the wires a few times for consistency; then calculate the pitch diameter:

$$PD = M - (3W - 0.866P)$$

where M is the micrometer measurement, W is the wire diameter, and P is pitch. The term "3W–0.866P" is the "constant" given in the charts that come with commercial wire sets.

I don't have a set of wires, but I do have a collection of small drills down to #75 (a closeout deal), three or more of most sizes. Figure 5-17 shows a 1/2-13 thread being measured using three #55 (0.0515" shank) drills. The micrometer reading was 0.540", which makes the pitch diameter 0.540 – (0.1545 – 0.866 x 0.077) = 0.452". This is a tad larger than the 0.450" Class 3 spec in Table 1, so if that's a concern, another pass with the screw cutting tool is called for.

FIGURE 5-17   Measuring the pitch diameter of a thread using three wires.

One tip on handling the wires: They are tiny, easily lost in the chip tray, so catch them with a sheet of paper. Masking tape was used in the photo in the figure to hold the pair of wires parallel, correctly separated. Others use modeling clay, even grease. Keep the wires (or drills) clean and straight.

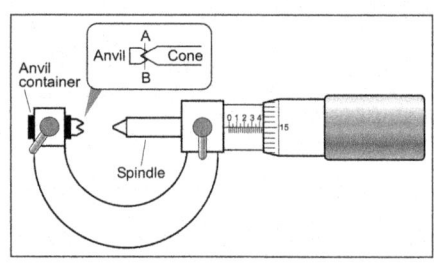

FIGURE 5-18 Screw thread micrometer schematic.

5-26 USING THE SCREW THREAD MICROMETER

Finally, we come to the screw thread micrometer (Figures 5-18 and 5-19), a lot easier to use than thread triangles or wires, but a lot more expensive.

FIGURE 5-19 Screw thread micrometer in use. This unit came with five anvil/cone pairs for thread pitches from 0.0156" to 0.2" (64 – 5 TPI), metric pitches 0.4 mm – 5.0 mm.

Name-brand thread micrometers go for $250 and up, some of them limited to only one or two thread pitches (*example:* 0.0333"/0.0357" for 30 and 28 TPI). Some imports (mine included) cost around $100. Mine came with a set of anvils and cones covering a range of thread pitches. If the micrometer has exchangeable components, it needs to be calibrated before use as follows:

1. Install the appropriate anvil/cone pair as shown in Figure 5-18.
2. Push the anvil container a few millimeters out toward the spindle; then *lightly* lock the container.
3. Run the spindle in so that the cone touches the anvil (line AB).
4. Slowly continue rotating the barrel, pushing back the anvil container, stopping when the barrel indication is exactly zero.
5. Fully lock the anvil container. Run the barrel back a few thousandths, then forward again to recheck zero.

Assuming the appropriate choice of anvil-cone pair, the pitch diameter can then be read directly if (and only if) the thread is an accurate 60°. There is always a small "tilt" error, usually ignored, because the Vee of the thread is offset from one side to the other, depending on the lead angle and major diameter. (The lead angle of a 20-TPI thread is about 4° for a 1/4" screw and only 1° for 1/2" with the same pitch.)

When using a screw thread micrometer, check that it is as nearly right-angled to the screw axis as the lead angle permits (in other words, check that the cone hasn't jumped a thread); and as with all micrometers—especially this type—take several measurements to be sure you are reading the largest dimension.

5-27 SOME OF THE FINER POINTS ABOUT SCREW THREADS

SCREW THREADS AT A GLANCE

1. Not all 1/4-20 screws are the same
2. How taps vary: The H limit
3. How threads fit: Unified (UN) thread classes
4. Metric taps are classified differently
5. Hole sizes for tapping

1. Not All 1/4-20 Screws Are the Same

Ever tried measuring your screw collection? Not the whole lot of them, of course, just a sampling. My various 1/4-20 screws have major diameters ranging from 0.243" to a little shy of 0.250". The #10-32 screws measure from 0.186" to 0.188", none of them the nominal 0.190". Ditto the #8-32 sampling: nominal 0.164", actual 0.150" to 0.162". But for all the variation, they do the job. In other words, they fit the threads I cut using taps, as well as practically any nuts chosen at random.

So, it seems we put up with a lot of variability, seldom giving it a second thought. When you do give it a second thought, the 200-plus pages on screw threads in *Machinery's Handbook* might make you wonder how threads actually get to be cut at all. But here's the thing: The vast majority of screw threads in the model shop are cut on the "A fits B" principle. Unless you are working to specs, measurement doesn't come into it; the job is done if the nut runs smoothly on the thread.

Sometimes, though, we need to look below the surface. For instance, you might wonder why a nut threaded with tap X fits, but one threaded with tap Y does not.

2. How Taps Vary: The H Limit

Taps X and Y may look the same, but chances are their *pitch diameters* are different. Leaving that aside for the moment, we'll assume they are both high-speed (HS) steel taps, with ground threads (G). If there is no "HS" mark, the tap is carbon steel, aka "tool steel," possibly with cut threads—not ground. (*Golden rule:* If you don't see "HS" or "HSS" on a tool or its packaging, assume the worst—it won't last, and it will soften if overheated.)

The following notes apply only to ground HS taps, HSG.

Now get ready for some tricky definitions. For instance, another thing you will see on an HS tap is an H number, the H meaning *higher than the basic pitch diameter* (there is also "L1" signifying smaller than basic, a rare item). H numbers range from H1 to H7, the higher the larger. For instance, a #10-32 tap marked "H3" has a pitch diameter 0.001" to 0.0015" over basic. This is where "the books" pile it on, for those still listening . . .

An H3 tap cuts a Class 2B thread. "Class 2" means a middle-of-the-road "standard fit," a good choice for most of the things we do. The suffix B defines this as an internal thread. External threads have the suffix A.

3. How Threads Fit: Unified (UN) Thread Classes

The three UN thread classes are:

- 1 = loose fitting
- 2 = standard
- 3 = tight fitting

Now, for even more confusion:

You might think that an H3 tap in other sizes will cut the same 2B class of thread. Not so! For a 2B #4-40 thread, you need an H2 tap; H3 for #8-32; H4 for 1/4-28; H5 for 1/2-20. Don't look for the logic in this—it is what it is.

My own collection of taps is all over the place. They came from all over the place, too—acquired from closeouts, garage sales, wherever. They remain unsorted by H limit, which explains why I don't just use the first tap out of the box. If you are working to tight specs, be sure when buying taps, you can specify your choice of H limit. Some suppliers don't know what they have in the stockroom—a tap is a tap.

The H limit applies only to the Unified thread series, coarse and fine (UNC example #10-24; UNF example #10-32).

4. Metric Taps Are Classified Differently

For metric taps the size limit is stated as a D number: This indicates the number of 0.0127-mm (0.0005") steps over the basic pitch diameter, as in M3 x 0.5 HSG D3 (3-mm major diameter, 0.5-mm pitch, ground high speed, 3 x 0.0127 mm over basic PD). The good news about Unified and metric threads is they both use the same 60° thread profile.

5. Hole Sizes for Tapping

One last point about taps: Tables of recommended drill sizes don't say so, but they are designed to give you a 75% depth of thread (full depth being the difference between major and minor diameters).

Do you really need that strong a thread? Reliable sources say there is no increase in strength over 60% of the depth of thread, which means you can drill one or two number sizes larger than the table, and use a lot less effort winding in the tap (less chance of breakage, too).

CHAPTER 6

Machinist's Precision Level

CONTENTS AT A GLANCE

6-1 OFF-THE-SHELF LEVELS

Practically all machine tools, especially lathes and milling machines, have to be carefully leveled to perform as expected. Off-the-shelf levels from the hardware store are good only for a preliminary setup, but they're a lot better than nothing.

6-2 PRECISION LEVELS

Precision levels are in a different class altogether. Key differences between precision levels and the hardware store variety are:

- **Cost.** Between 10 and 20 times more for comparable lengths.
- **Weight.** Precision levels have heavy bases, sometimes cast iron.
- **Sensitivity.** A precision level can typically detect as little as a 0.001" shift in height at one end relative to the other.
- **Adjustability.** Precision levels can be set to read zero when resting on a truly level surface.
- **Settling time.** It can take as long as 20 seconds for a precision level, but no time at all for the ordinary level.
- **Fragility.** Construction-style levels withstand a fair amount of abuse. Not so the precision level—if dropped, the vials may shatter, or the precision ground base may be damaged.

The following notes describe a representative precision level: A typical precision level will have a large longitudinal bubble and a smaller bubble at right angles to it. The smaller bubble is used for rough leveling. Some levels are accurately zeroed as shipped from the factory, in which case thread locker or varnish will have been applied to the adjustment screw(s). The level described here is more typical in that some degree of fine-tuning is expected. Even if yours is a pre-zeroed level, thorough checking is a good idea; if necessary, break the screw seal, adjust, and then reseal.

6-3 ZEROING THE LEVEL

You do not need a truly level surface for this, but you do need a *hard, flat surface* that can be set approximately level on the bench. Suitable surfaces are thick plate glass, or floor tiles of marble or ceramic. Check with a straightedge that your test surface is really flat and that the level sits firmly—no rocking—in the working area. The photos in Figures 6-1, 6-2, and 6-3 show a 12" x 12" ceramic tile purchased from the leftover pile at Lowe's.

FIGURE 6-1 Leveling the tile on the bench. In this example, the tile rests on a rod (*left arrow*) and a plastic shim (*right arrow*).

For these photos, the test surface was leveled with a plastic shim and a metal rod resting in grooves on the bench top mat. True the surface using a toolbox level (Figure 6-2).

Select a working area for the precision level, marking the area with a fiber-tip pen. For better repeatability—important—tape a guide plate at the front edge (Figure 6-3).

Taped to the tile surface (Figure 6-3) are (1) a guide plate of scrap metal and (2) a shim to raise one end of the precision level relative to the other. The example in the figure shows the left end shimmed, meaning that the left side of the test surface was slightly lower than "true." The shim thickness, in this example 0.015", was determined by experiment.

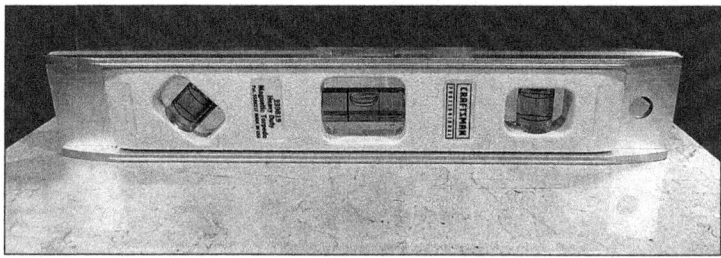

FIGURE 6-2 Rough-leveling the test surface.

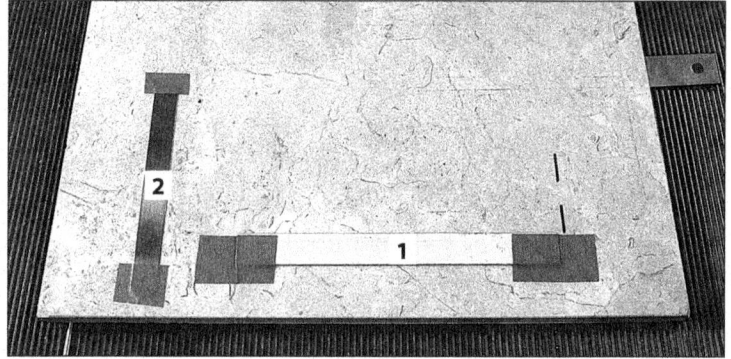

FIGURE 6-3 Preparing the test surface.

This may not be necessary if the test surface has been rough-leveled within a few thousandths of true across the span of the precision level.

You will know whether shims are called for when the level is placed in the marked area (Figure 6-4). If the bubble is "pegged" at either end, use shims to bring the bubble fully into the window, moving freely from side to side. Press gently down on the test surface to one side of the level to see which end needs shimming.

Allow 15 to 20 seconds for the bubble to settle.

Individual metal shims are best for this. Feeler gauges can be used only if they can be separated from the set to lie truly flat (no tilting from the hinge). If you have no metal shims on hand, consider using hard plastic such as a binder cover, even coated paper, anything that won't be compressed noticeably by repeated placements of the level.

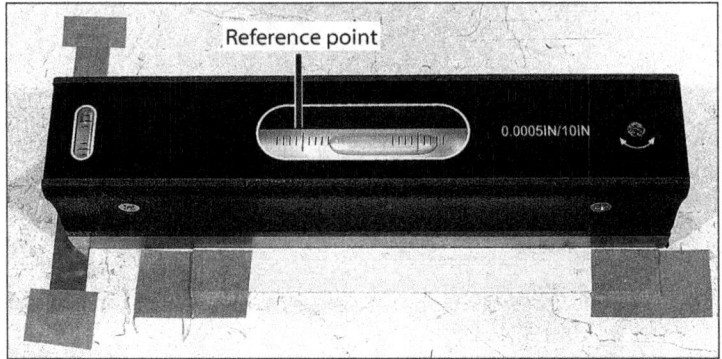

FIGURE 6-4 Shimming to bring the bubble into the window. The actual location of the bubble is not important, provided it is not pegged at either end.

With the level shimmed as necessary, proceed as follows:

1. Note the location of the bubble. In Figure 6-4, its left-hand edge is four divisions right of the reference point. Rotate the level 180° degrees, *reposition it exactly in the working area*, and then look at the bubble again after *allowing it to settle*. If the level is truly zeroed, unlikely at the first try, the bubble will be in the same location relative to the scale reference (Figure 6-5).

2. If the bubble is not in the same location, rotate the level to the start position; then make a *very small adjustment* to the leveling screw. In this example (shown in Figure 6-4), the screw is at the right end of the faceplate. Note the bubble's new location.

3. Rotate the level 180°; then recheck the bubble.

4. Repeat the process until the bubble is in the same location in both positions, within ± one-half of a scale division. (Your precision level may be specified differently.)

Another way to check accuracy is to *slowly rotate* the level on the test surface, away from the starting location, to bring the bubble to *the exact center of the scale.* With a fiber-tip pen or tape, mark the new location on the test surface (Figure 6-6). Rotate the scale 180°, and then set it down exactly in the new location. The bubble should again be centered.

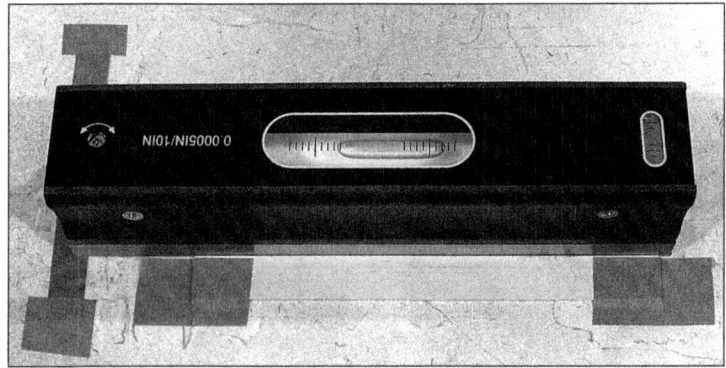

FIGURE 6-5 Rotate the level through 180°.

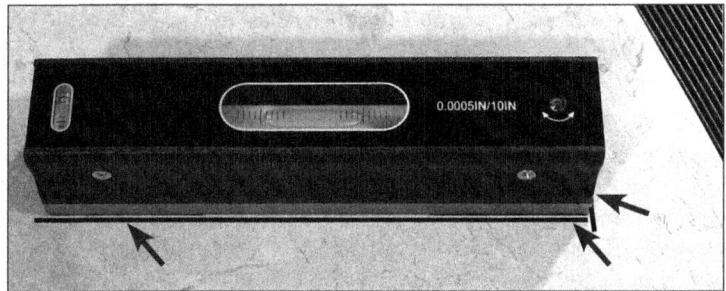

FIGURE 6-6 Does the bubble center properly? Mark the new test area (pointed to by the arrows) with a fiber-tip pen.

6-4 LEVEL ADJUSTMENT

All precision levels are similar in principle. The main bubble carrier is pivoted at one end, and is vertically adjustable at the other. The means of adjustment vary widely. Two examples are shown in Figure 6-7.

In the model shown in Figures 6-4 through 6-6, a Philips screw is typically concealed by a plastic plug. The screw runs in a tapped hole at the end of the bubble carrier, movement of which is stiffened by a compression spring, Figure 6-7, left. In another version of a similar level, practically identical in outer appearance, the rotatable screw is replaced by a fixed screw and two threaded circular collars on either side of the bubble carrier, Figure 6-7, right.

FIGURE 6-7   In the precision level at left, a compression spring (arrow) presses on the outer end of the bubble carrier. In the level at right, there is no stiffening spring. Instead, both collars must be tight against the bubble carrier. Adjust the collars using the key included with the level, inset.

Regardless of your level's adjustment mechanism, start by making very small changes, each followed by 20 seconds of settling time.

CHAPTER 7

The Self-Centering Chuck

CONTENTS AT A GLANCE

7-1 ABOUT THE SELF-CENTERING CHUCK

This is the workpiece holder of choice in most small engineering shops, so much so that in many cases it is almost a permanent fixture on the lathe. It is a classic example of a deceptively simple-looking device. The remarkable fact about self-centering chucks (aka "scroll chucks") is that most of them—even the budget variety—can hold a circular (or hexagonal*) workpiece with very small runout, often as little as a few thousandths of an inch.

The typical self-centering chuck has three jaws, but many variants are available. Here, we look first at the standard, very popular 3-jaw chuck (Figure 7-1).

FIGURE 7-1 Typical 3-jaw chuck on a 12" x 24" lathe. The arrow points to one of three bevel-gear sockets. They all turn in sync, but it is good practice to check for consistent tightness on all three before running the lathe.

---

*Hexagonal? Yes, this can be a great bonus when machining custom screws. Die-cutting screw threads puts a lot of slipping torque on the workpiece, often to the point where damaging force is used to tighten the chuck. A hex bar can be firmly held without undue strain on the chuck.

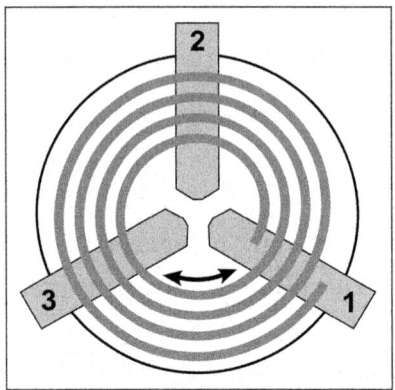

FIGURE 7-2   The jaws are moved in sync by a rotating scroll.

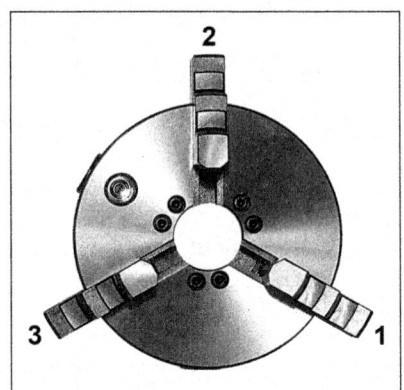

FIGURE 7-3   Internal jaws installed.

Self-centering chucks are operated by a precision-machined scroll (Figure 7-2). On the outer rim of the scroll's back surface is a bevel gear (not shown), which is turned by any of three bevel gear stub axles. All three internal gears move together when any one of them is turned using the square-tip wrench.

All 3-jaw chucks come with two sets of jaws, internal for smaller workpieces (Figure 7-3), external for larger. Unlike those in a 4-jaw independent chuck, self-centering jaws cannot be removed and flipped end for end. This is because the teeth on the back of each jaw are scroll-shaped—not plain screw threads—and can only mesh in one direction (Figure 7-4).

7-2 REMOVING THE JAWS

Before doing *anything* with the chuck, protect the bed with a wood scrap. Optionally, use a rubber band, as shown in Figure 7-5, to prevent the jaws dropping out. To remove the jaws, turn the square key counterclockwise.

The first jaw released by the scroll is #3, followed by #2 and finally #1.

A rubber band may not be necessary if you know which jaw is which—for instance, a fiber-tip pen can be used to number the chuck body. Given that information, all you need to do is test jaw #3 repeatedly while rotating the key counterclockwise.

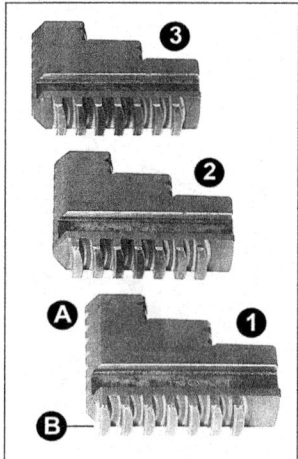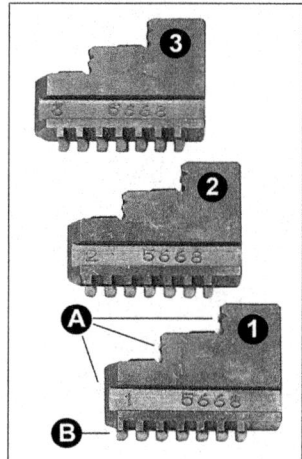

FIGURE 7-4 Internal jaws (*left*), external jaws (*right*). Note the separation between the gripping surface(s), A, and the inner tooth, B. The greater separation of the corresponding points in jaw 2 versus jaw 1 and in jaw 3 versus jaw 2 compensates for the difference in the scroll's radius 120° away.

FIGURE 7-5 Protect the lathe bed.

7-3 REINSTALLING THE JAWS

To reinstall the jaws:

1. Rotate the square key counterclockwise, backing the scroll until its outermost tip is just visible in slot #1 (see the inset in Figure 7-6).

2. Rotate it counterclockwise a little more to fully clear the slot; then insert jaw #1.

3. Rotate the key clockwise while checking that jaw #1 is now controlled by the scroll, moving inward to the center.

4. Continue rotating the key clockwise until the scroll's outer tip appears in slot #2.

5. Back the scroll to fully clear the slot; then insert jaw #2.

6. Repeat for slot #3, jaw #3.

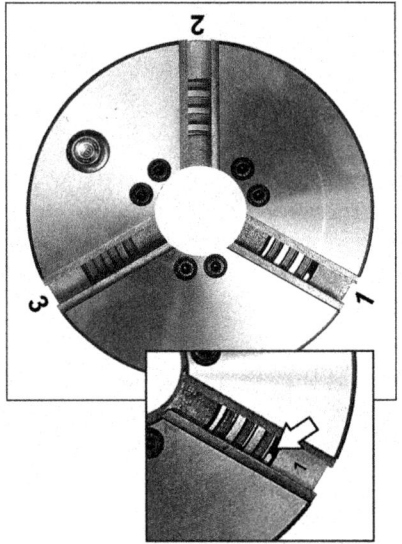

FIGURE 7-6 Reinstalling the jaws; the arrow points to the outermost tip.

Jaw locations are not optional! Each jaw is custom-fitted to its corresponding numbered slot.

If testing for runout using a dial indicator, be sure to use ground stock, such as a large-diameter drill rod.

(Nominally circular cold-rolled or other off-the-shelf rods may themselves be out of round by as much as 0.005".)

If the runout is unacceptable—or worse than expectations—this may be due to metal fragments on the jaws and/or the scroll. If so, this can be corrected by removing the jaws and cleaning the scroll with a brush inserted into the jaw slots.

Also bear in mind that runout can vary with workpiece diameter: For example, on a 1"-diameter test rod, the runout might be 0.003", reducing to 0.002" on a 2" rod—or the variance could go the other way. The main cause of this, aside from trapped *metal fragments* and *out-of-roundness* of the rod, is inaccuracy of the scroll. This is not uncommon, even with the highest-quality chucks, over time.

Are there ways of correcting runout in a 3 jaw?

Two suggestions to try:

1. With an inch or more of the workpiece forward of the chuck jaws, gently tighten the vise; then rotate the chuck slowly to locate the workpiece's indicated high spot. Using a nonmarring hammer (aluminum or brass), tap the high spot *lightly* inward toward the lathe centerline. Tighten some more, indicate, and then tap again if necessary. Sometimes this works wonders, sometimes not.
2. Insert 1- or 2-mil shims between the workpiece and jaws. This is in the same class as suggestion #1—sometimes you get lucky.

It may also be possible to true the gripping surfaces of the jaws using a tool-post grinder, but this calls for a commercial-grade grinder, not a shop-made unit like the one described earlier in the book in Section 3-9.

7-4 A RELIABLE WAY TO CORRECT RUNOUT

Aside from the hopeful possibilities above, successful or not, there is a better way. Needless to say, it comes at a premium—think *$800 and up at 2022 prices*. Several manufacturers produce *adjustable runout chucks*, usually featuring the tag "Tru" in the catalog—Bison Set-Tru, Röhm Hi-Tru, etc. These are self-centering chucks that can deliver concentricity as near perfect as you can achieve with an independent 4 jaw, but they do it in seconds over a wide range of workpiece diameters (although that capability can be compromised over time, hastened by overtightening—exactly the same in that respect as a regular scroll chuck).

The underlying principle of the adjustable runout chuck is simple—a loose-fitting chuck body, snugged to a shoulder on the backplate with four set screws (Figure 7-7). (Another example of this type of adjustment, on collet chucks, is shown in Chapter 1, Figure 1-23.) The chuck body is secured to the backplate by three or more cap screws, loosened enough (snug) to allow adjustment, then retightened once the runout has been checked and minimized.

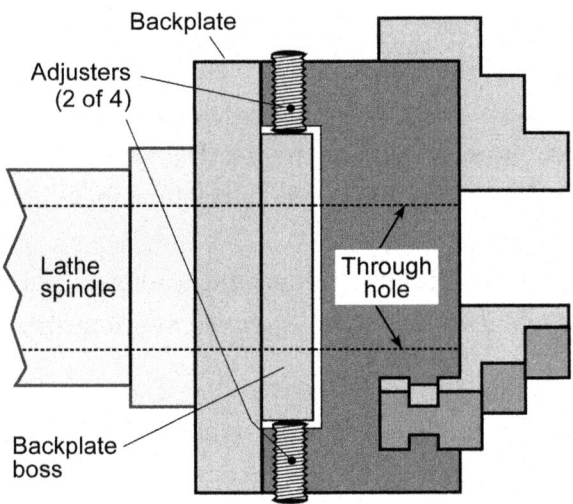

FIGURE 7-7 Adjustable runout chuck.

To adjust the chuck, install a truly round test piece of about the same diameter as the intended workpiece; then indicate and adjust using the four set screws (following the procedure recommended for the 4-jaw independent chuck in Chapter 4, Section 4-33). Fully tighten the cap screws and reindicate; if necessary, back off the cap screws and adjust again, looking for a TIR of maybe less than 0.0005".

Many users have found that a new chuck, once adjusted as above, can deliver an almost perfect TIR over a wide range of diameters, say, 1/2" to 2". If you are installing a new chuck, mark it so that it will install every time in the same location relative to the lathe spindle. Check runout of the backplate's outer ring—the surface that mates to the chuck—with a *left-facing* indicator: If there is any perceptible runout, correct it by skim-cutting. If you have to remove an appreciable amount, *remove the same amount* from the outer surface of the backplate boss to ensure that the chuck body can still mate solidly to the outer ring.

7-5 ANOTHER REFINEMENT

If your workpieces alternate routinely between large and small diameters, this can cause a significant amount of downtime for jaw swapping. This is tiresome enough to cause some machinists to consider other solutions, such as a second 3-jaw chuck with jaws installed "the other way." A more elegant fix, also expensive, is a chuck with *reversible jaws*, a feature sometimes available with adjustable runout chucks. The scroll mechanism in a "reversible chuck" is the same as in the standard model. So, too, are the teeth of the jaws that engage the scroll. The jaws themselves, however, are two-piece assemblies. The inner portion of the jaw, with scroll teeth as in Figure 7-4, is simply a *sliding carrier* for the outer portion (Figure 7-8). A crosswise slot in the carrier engages a key on the outer jaw to register the jaw precisely when it is flipped end for end—large workpiece to small workpiece and vice versa. This seems like

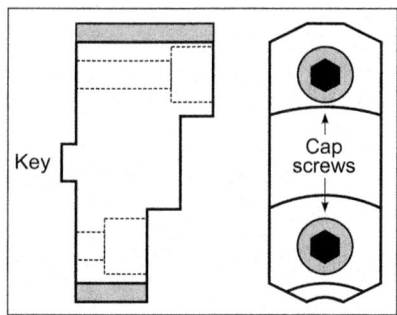

FIGURE 7-8 Outer portion of a reversible chuck jaw.

a neat alternative to repeatedly removing and reinstalling conventional chuck jaws, and so it is. However, there is often a need for considerable effort in removing the cap screws, then freeing the outer jaws from the carriers. Everything is necessarily very tight! This, together with the extra cost of reversibility, may suggest that the standard chuck, with its two sets of jaws, may not be so bad after all.

7-6 THE 6-JAW CHUCK

Aside from the obvious, the 6-jaw chuck (Figure 7-9) is similar in all respects to the 3-jaw chuck. It is a self-centering scroll chuck with exactly the same options—adjustable runout and reversible jaws.

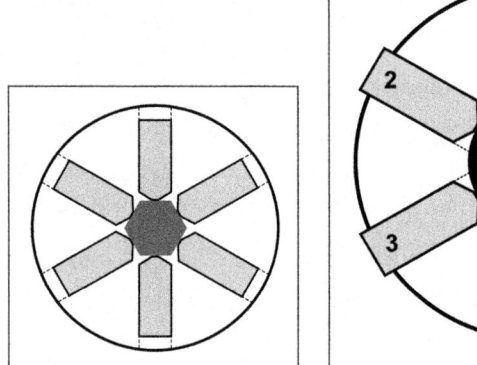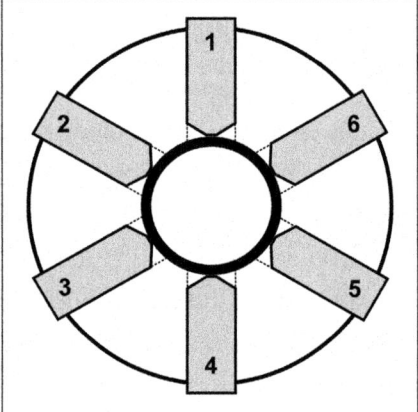

FIGURE 7-9   A 6-jaw self-centering chuck. Shown here (right) with a thin-wall workpiece. Just like a 3-jaw chuck, the 6-jaw can be used with hexagonal stock (*left*).

Size for size, a 6 jaw costs as much as 75% more than a 3 jaw from the same manufacturer. This usually puts it out of range for the small shop—unless, that is, you do a lot of work on thin-wall tubing. Here, the 6 jaw can make the difference between easy to do and impossible. Twice the number of jaws, evenly distributed around the workpiece, means extra gripping force without the distortion you tend to get with a 3 jaw.

7-7 LONG MATERIAL

The following applies to long material held in any type of chuck, self-centering or independent. If the project calls for a number of identical parts to be machined from bar stock, it is often convenient to start with material longer than the few inches or so we normally use. However, the question is: *What to do if you have on hand two or three feet of the material and don't want to cut it (wastefully) into shorter lengths?* It cannot simply be held in the chuck with only the work portion outstanding, because its tail will gyrate uncontrollably—partly overcoming the chuck's gripping force—causing similar, smaller oscillations at the machining end.

To machine long material, consider installing a *spider* on the left end of the lathe's hollow spindle. (Who knows why "spider" is the word for it.) This is available as an accessory for some lathes, but not many. It is typically an internally threaded cylinder that screws onto the left-hand end of the spindle. On the outer end of the cylinder are four brass clamp screws at 90° spacing. Just as on a 4-jaw chuck, these are adjusted to hold the bar stock concentric with the spindle.

Making a spider of the conventional sort calls for internally threading a thick-wall bushing, ideally steel. There's nothing unusual about that, except for the need to make sure it will fit the spindle while it's *still in the chuck*—almost impossible. One way to do that is to make an interim GO/NO-GO part, shown earlier in Chapter 4, Figure 4-12. The other way is to cut the thread so deeply that it's bound to fit. Or forget the threading altogether—instead, make a *flanged bushing* that's a push fit in the spindle (Figure 7-10).

The flanged bushing cannot accidentally detach itself from the spindle, provided the bar stock is firmly clamped at one end by the screws on the spider, and at the other end by the chuck. To advance the stock, simply ease the clamp screws and the chuck. Then, re-tighten firmly. If you will be working mostly with round or hexagonal stock, drill and tap a group of three holes at 120° spacing, instead of (or in addition to) the four holes at 90° spacing shown in Figure 7-10.

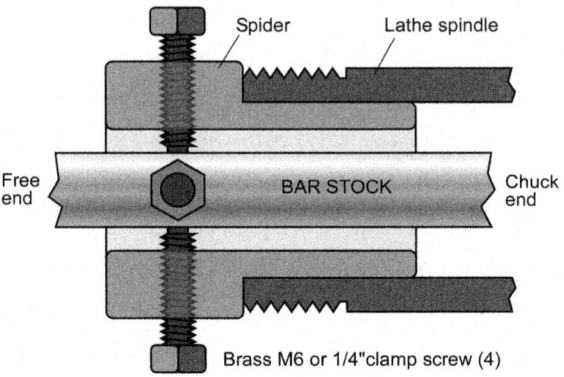

FIGURE 7-10   Simplified lathe spider.

Knurling

CONTENTS AT A GLANCE

8-1 AN OVERVIEW

Knurling is the process of impressing decorative or grippable patterns, usually diamonds, onto the surface of a round workpiece turned in the lathe. Knurling is also used to build up the diameter of a damaged shaft for a better fit in a ball bearing or sleeve. It is a deforming process, not a cutting process, calling for unfamiliar, even mysterious procedures. Even experienced machinists look on knurling as hit or miss—sometimes it works fine, sometimes it doesn't, for nonobvious reasons. The following notes explain the likely causes of difficulty.

Two key points to bear in mind:

1. The relationship between workpiece diameter and the knurl pitch is important (see Figure 8-7, later in the chapter). Don't take a chance on the first knurl wheel that comes out of the box. Measure!
2. On a complex workpiece with threading, boring, etc., do the knurling first if you can. That way, you won't have to scrap the piece and start over if the knurling doesn't satisfy.

Knurls are typically high-speed steel (HSS) wheels with sharp teeth, either "straight," in line with the wheel axis, or "diagonal," at 30° to the axis (sometimes 45°). Straight knurls are often used to impress a pattern around the rim of an adjusting nut; also, they are frequently used on small-diameter shafts to stake soft metal or plastic components.

Diagonal knurls—the most popular by far—produce a diamond pattern when used in pairs, left hand (LH) and right hand (RH). They can be *traversed along the workpiece to create any length of pattern.*

8-2 KNURL HOLDERS

In many applications, especially on heavier machines, the knurl wheels are pushed directly into the workpiece by the cross-slide or compound, using a holder of the type shown in Figure 8-1. This can result in an unacceptable load on the workpiece, chuck, and other components. Because of this, diamond knurling is most often done using opposing diagonal knurls that are clamped across the workpiece in a floating or scissors-style holder (Figure 8-2 and 8-3).

Push-style knurling can result in an unacceptable load on the workpiece, chuck, and other components. Because of this, diamond knurling is often done—especially in the small shop—using opposing diagonal knurls that are clamped across the workpiece in a floating or scissors-style holder (Figures 8-2 and 8-3).

Diagonal knurls are the only type that can be used in pairs! All other knurls must be used singly.

A one-piece diamond-pattern knurl delivers the same pattern as the knurl pairs on the previous page, but with the downside that a single wheel requires a lot of pressure. This is

FIGURE 8-1 Push-style double knurl. This is a typical knurl holder often supplied as part of a quick-change tool-post kit. The knurl wheels are usually not specified. In this example, they are 3/4" diameter by 3/8" wide, with 45° teeth at a 16-TPI (teeth per inch) circular pitch. Careful height setting is necessary for even pressure on both knurls.

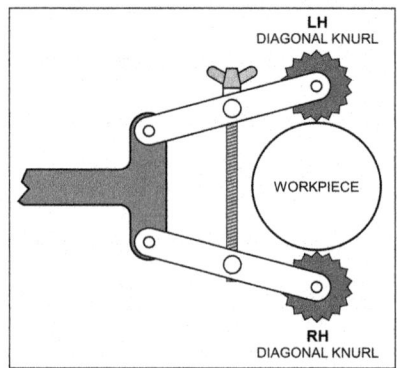

FIGURE 8-2 Floating clamp knurl holder schematic (see also Figure 8-3). The knurl wheels are clamped across the vertical diameter of the workpiece. Exact fore-aft positioning is not critical. Height setting is also unimportant, because the arms float freely up and down. This type of holder is ideal for small machines, because it eliminates most of the deflecting load.

why single knurls tend to be used on soft, ductile materials such as copper or aluminum.

FIGURE 8-3 Shop-made holder for diagonal knurls. This is a floating-style holder designed for 5/8"-diameter, 1/4"-wide knurl wheels. The knurl wheels are clamped above and below across the vertical diameter of the workpiece. Clamping pressure is applied by tightening the socket head cap screw (arrow). Holders like this need to be robust enough to withstand heavy compressive and sideways forces. Also, be sure the wheels can be easily removed to allow cleaning of their side faces.

8-3 KNURL WHEELS

Knurl wheels are available in many diameters and widths, with various tooth patterns and axle sizes (Figure 8-4). Aside from wheel size, the main choice is circular pitch (CP), the number of teeth per inch around the knurl's circumference (Figure 8-5). Many small shops often get by with just one knurl setup for every project: For instance, this might be a LH/RH pair of 16 teeth-per-inch knurls, often regarded as coarse. By that reckoning, 20 and 30 teeth per inch are, respectively, *medium* and *fine knurls*. But there are dozens of other choices, so it comes down entirely to personal preferences.

Also, please note that catalog descriptions of knurl wheels are often vague and incomplete. When ordering for a specific requirement, e.g., plain face or chamfered, you need to ask.

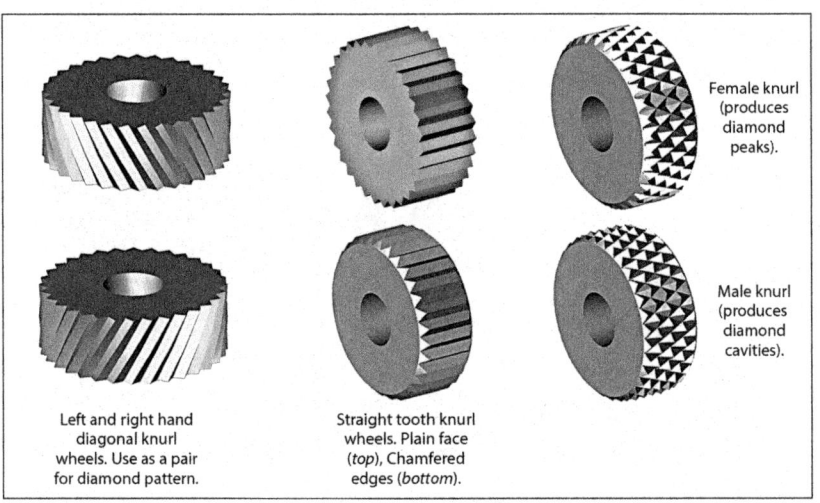

Left and right hand diagonal knurl wheels. Use as a pair for diamond pattern.

Straight tooth knurl wheels. Plain face (*top*), Chamfered edges (*bottom*).

Female knurl (produces diamond peaks).

Male knurl (produces diamond cavities).

FIGURE 8-4 Knurl wheels are available in numerous sizes with various tooth patterns. Diameters range from 1/2" to 1", widths from 3/16" to 3/8". The hole diameter is usually 1/4". They are usually HSS, sometimes with TiN (titanium nitride) or other coatings. Most are available with chamfered edges for easier tracking along lengthy knurls. Other tooth forms, not shown, include convex and concave. (Convex-toothed wheels are recommended for high-volume production of long knurls.)

8-4 FIRST, MEASURE
THE KNURL WHEEL

Assuming that you don't know the specifics of the knurl you plan to use, the first job is to measure the wheel's circular pitch. One way is to count the number of teeth and then divide the knurl's circumference (π x diameter)

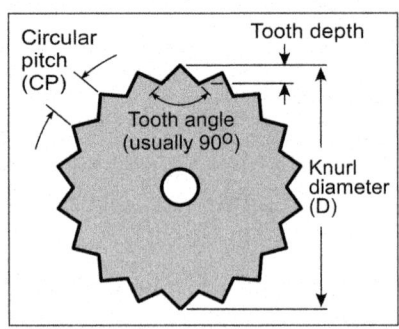

FIGURE 8-5 Knurl wheel dimensions.

by the tooth count. Or you can "fingerprint" the knurl by rolling it heavily onto any surface that will take an impression—soft wood, such as a paint stirring stick. The longer the print the better, at least 100 lines, marking every five lines for easier counting (Figure 8-6).

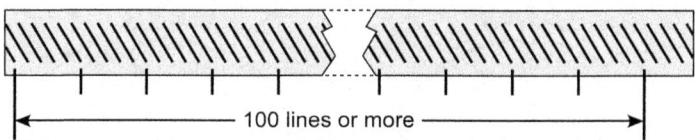

← ——————— 100 lines or more ——————— →

FIGURE 8-6 Knurl wheel fingerprint.

The fingerprint of the knurls I used for Figures 8-7 and 8-9 measured 5.84" over 100 lines. This gives us 17.12 teeth per inch around the knurl's circumference, so its approximate circular pitch (CP) is $1" \div 17.12 = 0.0584"$.

For good results the circumference of the workpiece should be a *whole-number multiple* of the measured CP, with no decimal places. If it isn't a whole-number multiple, the result will likely be "double tracking," an instantly recognizable problem caused by out-of-sync impressions on successive revolutions of the workpiece (Figure 8-7). If the knurl pattern seems to be finer than you would expect looking at the knurl wheel, you have double tracking (Figure 8-8).

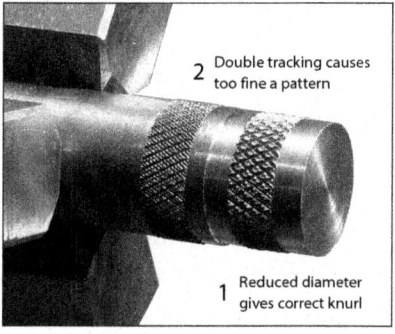

2 Double tracking causes too fine a pattern

1 Reduced diameter gives correct knurl

FIGURE 8-7   Good and bad knurls. Knurl 2 shows what can happen when you work without calculations—double tracking. For knurl 1, the workpiece diameter has been trimmed so that the circumference matches the knurl wheel's pitch.

In diagram 1, the workpiece has just started its first revolution, imprinting marks 0, 1, 2, and 3. Diagram 2 shows the workpiece almost full circle, at mark #21. For a perfect pattern, the next mark made by the knurl, indicated by the arrow, should coincide with mark #0. Here it's way off—and

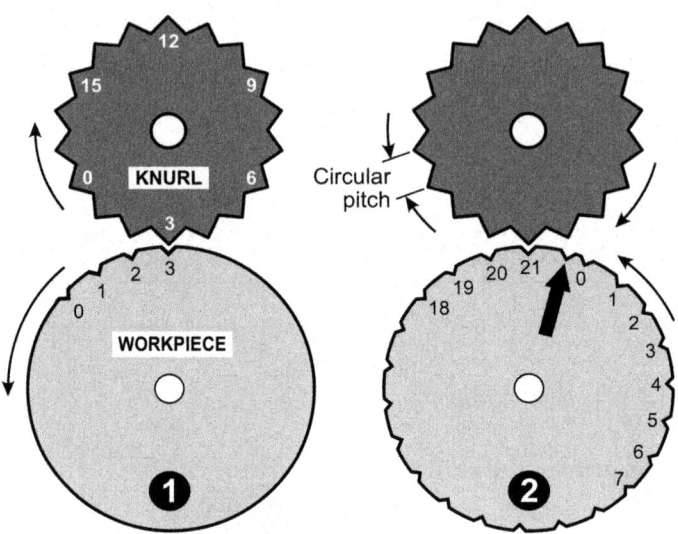

FIGURE 8-8   Why double tracking happens.

that will apply to other first-revolution marks around the circle. The fix for "double tracking" like this is to make the circumference of the workpiece a whole-number multiple of the knurl's circular pitch.

Figure 8-8 shows that it's important to match the circumference of the workpiece to the knurl's circular pitch, CP. But back in the shop, we think about diameter, not circumference.

So, the question is this: *How can we relate circular pitch to diameter, instead of circumference?*

The answer is easy: Divide CP by π. The resulting knurl factor can be used directly to calculate matching workpiece diameters. For the knurls in Figures 8-7 and 8-9, the factor is 0.0584"/π = 0.0186".

With this knurl factor, what can we expect if the workpiece diameter is 0.75"?

We get 0.75" ÷ 0.0186" = 40.32. Not good, because 40.32 isn't a whole number.

The nearest lower whole number is 40: Plugging in 40, we get 40 x 0.0186" = 0.744". This suggests that for good results we need to trim 0.006" off the diameter.

There is no guarantee of a perfect knurl with 0.744" as the starting value, but chances are much better than hoping for the best with no calculations. Starting pressure is also important. It is quite possible, even likely, that light pressure may show double tracking—but that may disappear with greater knurl pressure. Why? Because increased pressure reduces the effective diameter of the workpiece.

Here are a few more examples:

- **1/2" diameter.** 0.5" ÷ 0.0186" = 26.88. This is close to 27, which gives 27 x 0.0186 = 0.502", so probably OK without trimming.
- **14-mm diameter.** 0.551" ÷ 0.0186" = 29.63. The nearest whole number below this is 29, which gives 29 x 0.0186 = 0.539", so trim 0.012" off the diameter.
- **5/8" diameter.** 0.625' ÷ 0.0186" = 33.6". Trim 0.011" off the diameter (33 x 0.0186 = 0.614").

When planning a job, consider experimenting with an oversize scrap workpiece instead of thinning the diameter up front.

Note that in the above examples we are reducing the diameter by only very small amounts, just a few thousandths of an inch. It doesn't take much, so take care not to overdo it.

Material hardness is another factor to keep in mind: Workpieces of the same diameter in, for example, steel and aluminum respond quite differently to the same knurl pressure. Another factor to keep in mind is that the effective diameter of the workpiece decreases significantly as the knurl pattern deepens.

DIAMETRAL PITCH KNURLS

These is a distinct class of knurls (aka "diametrical pitch") that are pre-sized in manufacture to track properly—no double tracking—when used on fractional-size workpiece diameters up to 1", usually in 1/32" increments. They are manufactured to closer tolerances than the more commonly available circular pitch knurls and therefore tend to be more expensive. Diametral pitch knurls are available from only a few specialty suppliers.

8-5 KNURLING TIPS

- Use a low spindle speed, 150 to 200 rpm.
- Diagonal knurl pairs are usually started at the right end of the workpiece, with the knurls "hanging out" by about one-half of their width.
- With the saddle stationary, start by applying enough pressure to impress a definite pattern when you spin the workpiece by hand. If you see double tracking, and the material is soft, it may be that a little more pressure will deliver a correct, full-pitch knurl. The workpiece diameter may need to be trimmed if the material is harder.
- Run the spindle; then slowly feed the saddle by hand to the left. When the knurl is at the desired end point, leave the spindle running and apply more knurl pressure for the return pass. Repeat as necessary. Do not allow the knurls to become fully disengaged from the work until the job is finished.
- A knurl calling for multiple passes will be degraded if particles shed from the work are not flushed away continuously with an air line or a water-soluble coolant. If a pressure coolant supply is not available, a spray bottle can be used instead. WD-40 is also a good alternative flushing fluid. It can also

help if you clean the work while knurling using a small wire brush.

- Power feed can be used instead of hand feeding, but be wary of backlash issues. If the knurl is deep enough, chances are the knurls will stay in sync on the return pass. If in doubt, back the knurl to the starting point using the saddle hand-wheel without changing the spindle direction.
- Traversing the saddle to produce a long knurl (Figure 8-9) applies a heavy *swiveling force* on the tool post. Be sure the compound and tool post are securely fastened.

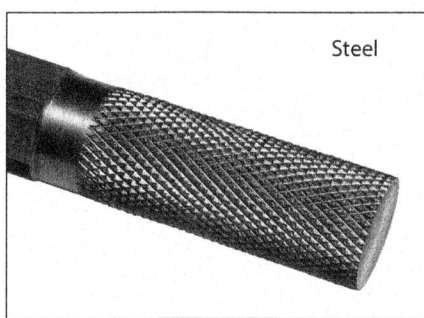
Steel

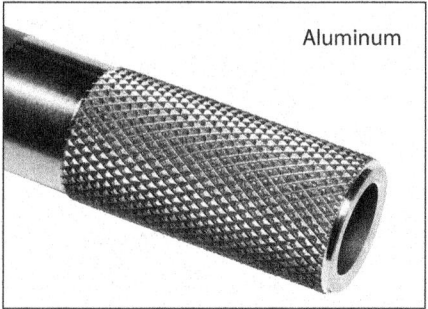
Aluminum

FIGURE 8-9 Long knurl examples. Both of these knurls were done using plain-face (not chamfered) knurls in the floating-style holder of Figure 8-2 (chamfered-edge knurls would have taken less effort to traverse, but were not available until I ground them, as in Figure 8-10).

- For a clean, crisp appearance, clean up the knurl with a fine file.
- Time-saver for diagonal knurl pairs—instead of clamping them across the full vertical diameter, pull the cross-slide back a few thousandths; then clamp across the shorter span. Now you can increase knurl pressure for successive passes simply by pushing in the cross-slide.
- If you are working on steel, and have a choice, use 12L14 or other free-machining steel instead of 1018. Bear in mind that standard HSS knurls do not work well on stainless steel.

- For easier traversing, plain-face knurl wheels can be chamfered by spinning them against a belt sander (Figure 8-10).

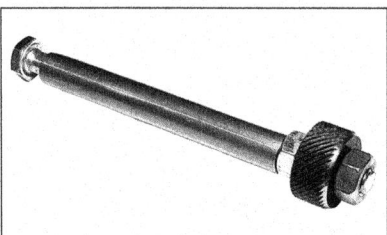

FIGURE 8-10   Holder for chamfering. Use something like this with a belt sander—a scrap of tubing and a long bolt. Hold the knurl at about 45°; there's no need for great accuracy.

HOW CAN KNURL PAIRS WORK?

We take a lot of care matching workpiece diameter to the knurl's circular pitch to avoid the problem of double tracking. So how can it be that a pair of *unsynchronized* LH and RH knurls above and below the workpiece don't cause the same problem? The bottom knurl is out of sight anyway, so you simply cannot know how the knurls relate to each other. Surprisingly, as this diagram shows, it hardly matters at all.

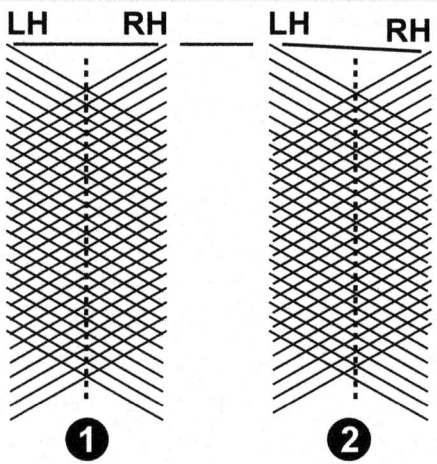

In column, 1 the top lines of LH and RH knurls are in perfect sync. In column 2, the top line of RH is displaced by about one-half of the knurl's circular pitch, but the diamond pattern is still perfectly formed. As shown by the red dashed lines, the central column of diamonds in column 1 has, in fact, shifted to the right by a fraction of an inch in column 2, and that's all.

Bear in mind this applies only to diagonal knurl pairs used for diamond knurling. You cannot use pairs of any other knurl pattern.

CHAPTER 9

Add Versatility by Indexing the Spindle

CONTENTS AT A GLANCE

9-1 UNUSUAL LATHE OPERATIONS

The idea of clicking the spindle around in exact angular increments was mentioned earlier in the book, in Section 3-6. Here is more detail on how you might do it and the many things you can do with it—for instance, drilling a series of through holes evenly around a flange or drilling radial holes for spokes on a wheel. Or even more unusual, hand-filing instead of machining flats on a workpiece to avoid having to take it away to the mill.

The conventional indexing solution—according to practically all lathe books for the past 100 years—is to attach a change gear to the left-hand end spindle. A 60-tooth gear, for example, gives these divisions: 2, 3, 4, 5, 6, 10, 12, 15, 30 and (of course) 60. For occasional use, the change gear is pretty adequate, but you will need to spend time working on a rock-solid, spring-loaded detent system.

Since most lathe spindles are hollow, one easy way to attach a gear is with an expanding plug (Figure 9-1, *left*). This method worked fine for me on two earlier lathes.

FIGURE 9-1   Spindle attachment options. *Left:* A hollow plug expanded by an internally threaded taper on a stub axle. *Right:* An internally threaded cap that screws onto the lathe spindle. The same threaded stub axle was used in both cases. The plain outer end of the stub axle locates the center of the dividing plate. Every lathe is a little different!

For lathe spindles with external screw threads visible, it may be more convenient to make an internally threaded end cap. This is what I did for my current lathe, which has a 1-1/2"-bore spindle (Figure 9-1, *right*).

9-2 WHEN THE NEED ARISES FOR MORE FLEXIBILITY

Detent and spring issues were enough for me to move on to a more elaborate setup with holes in a disc instead of teeth. What I have used—frequently—in the past few years is a 1/8"-thick, 6"-diameter steel "dividing plate" with three circles of holes giving 60, 36, and 16 divisions (Figure 9-2).

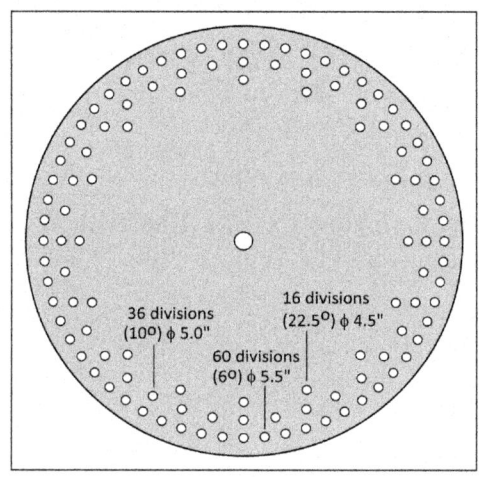

FIGURE 9-2   Shop-made dividing plate.

If you have a mill with a DRO, this is not difficult to make, just tedious—a few minutes listing sine and cosine values on a spreadsheet, followed by an hour or two drilling holes.

To go with the dividing plate is a 1/4"-thick steel indexing arm (Figure 9-3). The rectangular plate at the right of the arm serves as a socket for the stub-axle extension on the lathe spindle. It is adjustable to allow for inaccu-

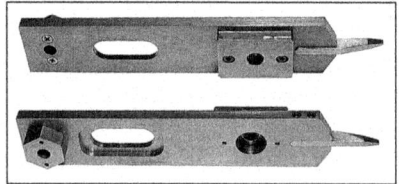

FIGURE 9-3   Indexing arm.

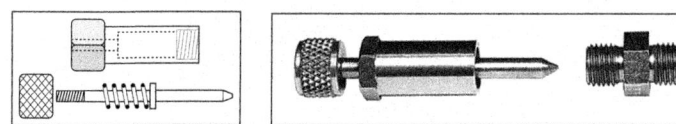

FIGURE 9-4 Spring-loaded indexing pin.

racies in attaching a mounting pillar to the cast-iron surface of the head-stock (shown later in Figure 9-6).

Other components of today's indexing setup are shown in Figure 9-4. Getting the spring pressure right took two or three tries—light enough to allow the pin to be pulled out for repositioning, yet strong enough to lock the plate solidly.

A recent addition is a vernier pointer that allows 1° increments using the outer ring of 60 holes (Figure 9-5). The pin for the vernier is a 1/8" dowel—no spring loading—on the same lines as a conventional spin indexer used on the mill.

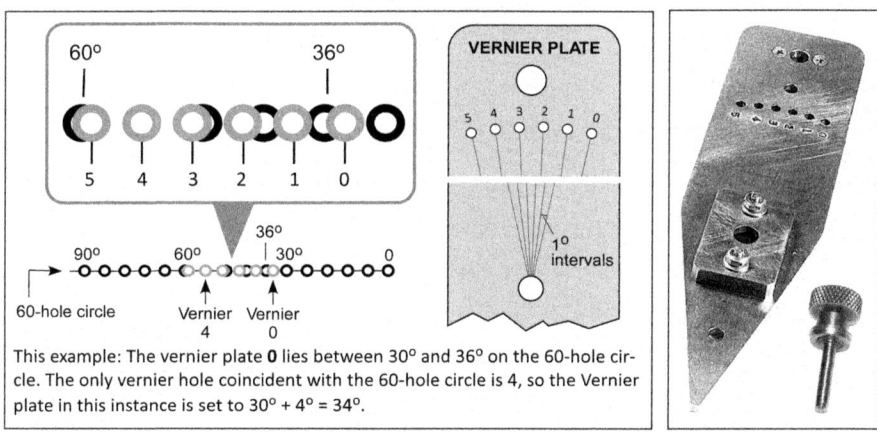

This example: The vernier plate **0** lies between 30° and 36° on the 60-hole circle. The only vernier hole coincident with the 60-hole circle is 4, so the Vernier plate in this instance is set to 30° + 4° = 34°.

FIGURE 9-5 Vernier indexing pointer. With this vernier pointer used in place of the basic indexing pointer, the outer 60 holes give 1° divisions, like a spin indexer on the mill.

Figure 9-6 shows the indexing assembly on my current lathe. The only part of this installation that caused extra concern was drilling and tapping for the support pillar—be sure to pick a "dead zone" in the headstock. (Take special care if your lathe has an *oil-filled* gearbox.)

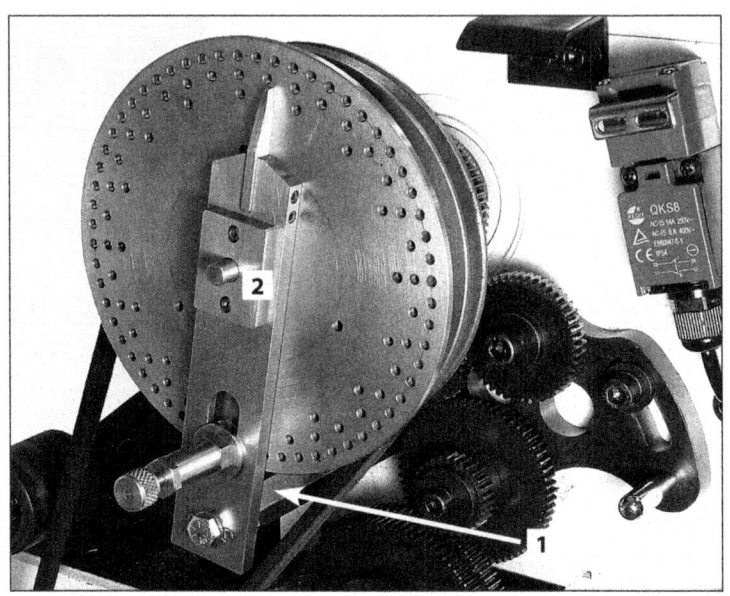

FIGURE 9-6 Complete indexing assembly. The indexing pointer is attached to a hex-section pillar (1). This needs extra-careful positioning to avoid interfering with other functions. The hub plate (2) allows a small amount of adjustability.

9-3 A QUICK-AND-EASY FILING REST

My first use for the spindle indexer, back in the 60-tooth change gear days, was with a filing rest that would make it possible to generate squares, hexagons, etc., without taking the turned piece from the lathe to the mill.

I came across this idea in an ancient lathe book (so ancient that it talked about treadle-powered lathes). Today, you can buy drawings and parts for a very fancy design, but it too belongs in the past. If your lathe has a quick-change tool post, you need nothing more than a couple of flanged rollers attached to a square bar (Figures 9-7 and 9-8).

FIGURE 9-7 QCTP filing rest.

The filing rest is surprisingly easy to use—fast, too. As you can see in Figure 9-9, there's not much to it—two case-hardened 3/4" rollers with flanges to keep the file on track. The flanges are 1" apart, wider than any of my files. The rollers run on 5/16" x 1" shoulder bolts in a 1/2" x 1/2" bar. The only thing to be careful of is the fit of roller on bolt; you need a true-running 3/4" surface, with only a few mil of axial side play—so be careful with the counterbore depth. The shoulder bolts are available singly from McMaster Carr, 91259A583. Because they can be undersize in diameter and oversize in length, you need to have them on hand before machining the rollers. As drawn, the rollers are 3" on centers, allowing better visibility than the smaller spacing shown in the photos.

FIGURE 9-8 QCTP filing rest setup. The filing rest is clamped in a shop-made toolholder, with 3/8-24 height adjustment, about 0.01" per 1/4 turn of the knurled disc. This is almost the same pitch as standard toolholders with an M10 x 1 thread.

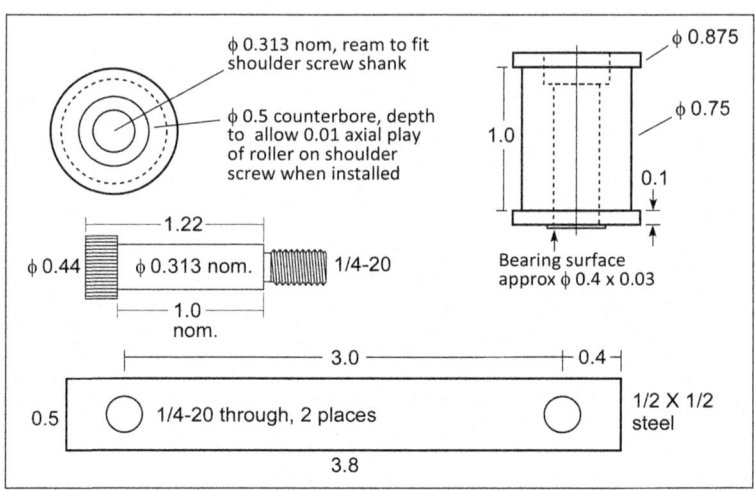

FIGURE 9-9 QCTP filing rest dimensions.

Figure 9-10 shows how the rest was originally intended to be used, with both rollers on the operator's side of the workpiece. Shown here are the first two flats of a "filed-in-place" hex head. The setup allows you to file at any location between chuck and tailstock, with or without a steady rest, on any diameter up to about 1".

FIGURE 9-10 Filing rest in front of the workpiece.

I realized after using the rest a couple of times that most of the filing I do is on the end of a workpiece, so the rollers can straddle the diameter (Figure 9-11). This is a bit more stable (but the other setup works fine, too). Figure 9-12 shows the filing rest setup for flats on a long workpiece—a job that's out of range for the typical milling machine, especially one equipped with the necessary means of indexing.

To use the filing rest, start by indexing the dividing plate to the desired location. Crank up the toolholder height until the file just grazes the upper surface of the workpiece. Lower the toolholder a half-turn or so on the knurled disc, about 0.02", lock the holder, and then file down to the roller depth. Index the dividing plate to the next location, and repeat.

FIGURE 9-11   Filing rest straddling the workpiece.

The only thing to watch out for is a forward tilt of the rollers due to inaccuracies in the toolholder, etc. The rollers need to be *parallel to the lathe axis*. Retighten the clamp screws, or shim if necessary.

FIGURE 9-12   Use a steady rest to file flats on a long workpiece. This allows the workpiece to be much longer than could be set up on a milling machine.

9-4 A TOOL-POST DRILL

This is another tool that's a real game changer, but very easy to make (Figure 9-13). What surprises me, looking back, is how long it took me to get around to it.

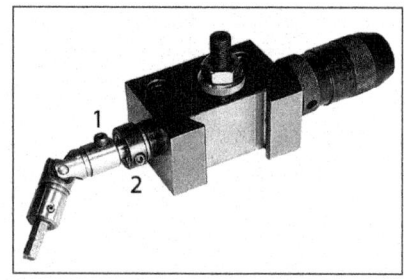

FIGURE 9-13   Tool-post drill. The socket head screw (1) and clamp collar (2) allow the assembly to be reversed for either longitudinal or radial drilling.

The starting point for the drill was a 100 Series "heavy-duty" QCTP 3/4"-diameter boring bar holder (Figure 9-14). I had two of these on the shelf.

The boring bar holder came with a full-width split-reducing sleeve, 5/8" ID, perfect for Oilite flanged bearings at each end. The chuck spindle was machined from 1/2"-diameter ground stock, probably 4140, turned down to 3/8" to leave a narrow flange at the drill chuck end. This was tapered Jacobs JT1 at one end to fit a Röhm 1/4" chuck. The hand-drill end of the shaft was turned to fit a universal joint—not

FIGURE 9-14   QCTP boring bar holder.

essential, but otherwise care is needed to keep the drill aligned. A 1/4" hex stub, installed in the outboard end of the universal, can be used with either a drill chuck or a standard hex driver.

9-5 SECOND THOUGHTS ON UNIVERSAL JOINTS

After the above was written, I discovered that universal joints like the one in Figures 9-13 and 9-19 are now hideously expensive. Also, I remember doing a fair amount of work in making a close-fitting 1/4" hex stub, etc.

The drawing in Figure 9-15 shows a different, much easier approach. The suggestion here is to use a ground 3/8"-diameter shaft, instead of machining down to 3/8" from 1/2". File a short length of 1/4" hex at one end, and taper the other end for the chuck. The larger JT1 dimension (0.384") is a tad larger than 3/8", but that is not likely to be an issue. Instead of the universal joint, drive the shaft using one of the many flexible drives on the market for less than $10.

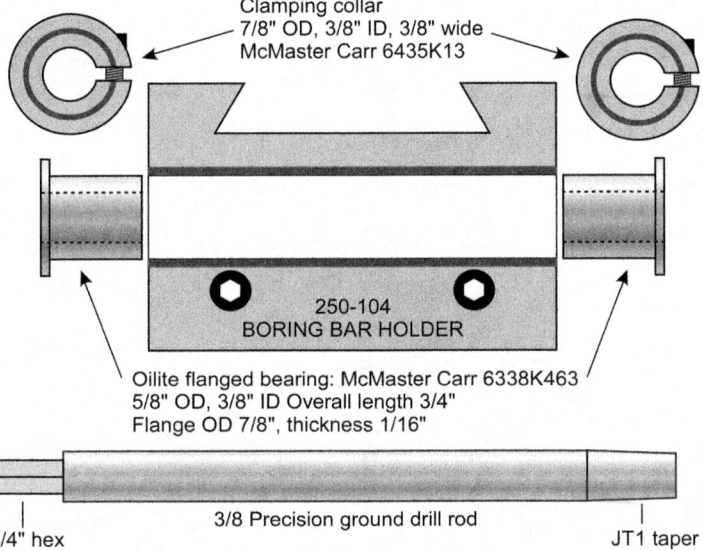

FIGURE 9-15 Modified tool-post drill. This version has no universal joint and is less expensive to make.

Two important refinements:

1. The socket head screws, shown above in Figure 9-13, allow the spindle assembly to be reversed in the QCTP holder, allowing it to be used both for longitudinal drilling (Figure 9-16) and for radial drilling (spokes on a wheel, set screws on a shaft—Figure 9-17).

FIGURE 9-16   Tool-post drill setup for longitudinal drilling.

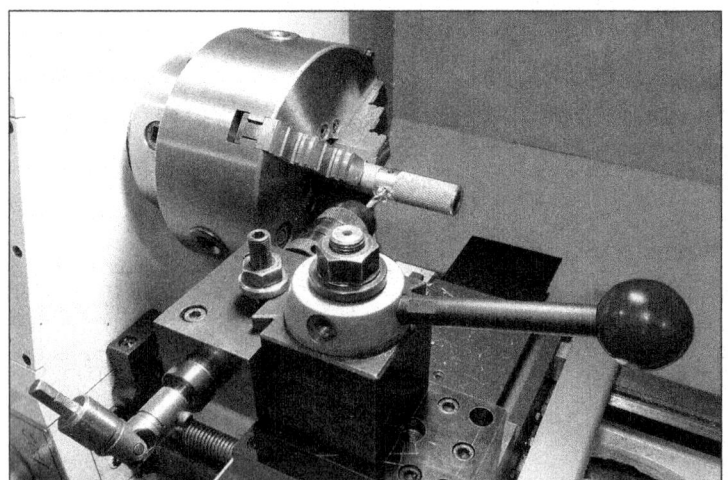

FIGURE 9-17   Tool-post drill setup for radial drilling.

2. The gap between the shaft flange (or "collar") and the Oilite bearing at the chuck end (Figure 9-18) is very useful when threading a hole you have just drilled. (This makes sure the tap is properly aligned—rarely achieved when attempted freehand.) Install the tap in the chuck; then apply light forward pressure with the saddle handwheel to seat the tap in

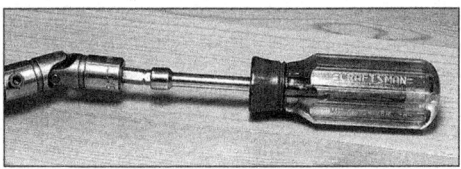

FIGURE 9-19   Hand-turn using a 1/4" socket wrench.

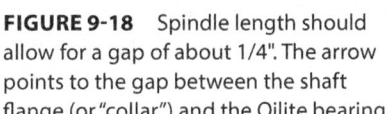

FIGURE 9-18   Spindle length should allow for a gap of about 1/4". The arrow points to the gap between the shaft flange (or "collar") and the Oilite bearing.

the hole. Now, while maintaining forward pressure, feed the tap in by hand-turning the shaft with a 1/4" socket wrench (Figure 9-19).

The tap (and the drill shaft) is thus pulled gently forward with no resistance from the saddle. When you don't need the gap, bring the clamp collar forward to seat against the rear flange, as shown earlier in Figure 9-13.

CHAPTER 10

Making a Tailstock Drill Press

CONTENTS AT A GLANCE

10-1 SOME BENEFITS

A tailstock drill press is a great time-saver when drilling small, deep holes, especially when there's a need for frequent withdrawal for oiling and cleaning. Additionally, lever operation gives a degree of sensitivity that's simply not possible with a lead-screw feed. For that same reason, the tailstock drill press is also very useful for feeding in small, fragile taps.

There is nothing new about the idea; Myford, a British lathe manufacturer, at one time offered a "Sensitive Drilling Adapter" like this, the earliest example I have seen in use. What's new here are specifics on constructing the device, and a couple of caveats on things that can lead you astray.

FIGURE 10-1 Tailstock drill press installed.

10-2 SOME SPECIFICS

Figure 10-1 shows the drill press installed on a 10" x 22" lathe with MT3 tailstock taper. There is nothing critical about the dimensions, which can be adapted for any tailstock taper from MT2 and up. From fully extended to fully retracted, the throw is approximately 2.7", more than adequate for most small-hole operations (so much so you may wish to shrink the dimensions for a more compact design). Access to a milling machine is a must. Suggested dimensions are given in Figure 10-4.

10-3 MT3 ADAPTER SLEEVE FOR THE TAILSTOCK

The project starts with an MT3-MT1 adapter sleeve, also known as a "taper extension socket." Look for "MT1 inside, MT3 outside, soft with hardened tang," about 6-7/8" overall length. This is available from several importers for around $20. Be sure that only the tang portion is hardened; otherwise, it will be impossible to machine.

Another possibility is a blank machinable arbor with an MT3 shank, about $10. The only issue with this might be a shorter machinable portion (only about 1-5/8" long), which will call for design modification. To machine the blank arbor, you will need another adapter sleeve, this one to go from your lathe's headstock taper, say from MT5 to MT3.

If your tailstock is MT2, the straight cylindrical portion of the adapter will likely be less than 0.82" diameter in the drawings, Figure 10-4, which may call for a smaller-diameter arbor.

10-4 ARBOR AND CHUCK

The arbor is 0.5"-diameter W1 or O1 ground-finish tool steel, 7.75" overall length (McMaster Carr 8890K1). You will need this on hand when you machine the adapter sleeve.

Start by cutting off the adapter sleeve tang, if it has one, using a hacksaw with an old blade. (You could use a cutoff wheel instead, but I found the blade to be more efficient.) Minus the tang, the adapter sleeve will be about 5.2" long overall, with the tapered portion a shade longer than 3". Length dimensions are not critical.

Set the cylindrical end of the adapter sleeve in a 4-jaw chuck, doing the best you can to minimize runout. Then, very carefully, step-drill it—in small increments—from the taper end. This is easy to say, tough to do, so much so you might want to have an extra sleeve on hand at the start, plus a handful of old drills. Finish the bore by reaming 0.0005" to 0.001" oversize to allow a smooth-sliding fit for the arbor.

If you have a headstock taper adapter, MT5 to MT3 in my case, you can neck-down the adapter sleeve's cylindrical end a few thousandths to provide a shoulder for the back stay. This is not an essential step, but bear in mind that this is the spot that takes most of the drilling load. It is easier to drill the frame member first, and then neck-down the sleeve to suit.

Cut the arbor to length, 7.75" overall; then mill an air relief slot 0.125" wide x 0.063" deep, long enough so that the end of it is just visible in the tang slot when the arbor is fully retracted. Be sure to drill the 3/16" hole (for the pivot pin) near the arbor nose on the same centerline as the air relief slot.

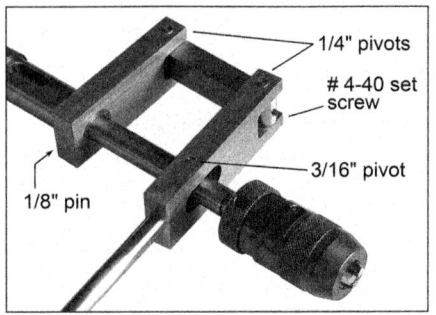

The 1/4" (6.5-mm) chuck in Figure 10-2 is a Röhm Supra, JT1 taper, which was good value at the time. Today, 10 years later, it goes for $100 and up. Similar products are available for much less, around

FIGURE 10-2   Tailstock drill press components.

$40: Look for a 1/32-1/4 JT1 precision keyless chuck. You might go with a threaded chuck instead, but bear in mind that taper socket chucks have less runout and better pointing accuracy. (I learned this the hard way years ago on an early version of the same design.)

10-5 MACHINING THE JT1 CHUCK TAPER

Figure 10-4 (1) shows a necked-down 0.375"-diameter portion behind the taper. This doesn't have to be done—it serves only as a visual demarcation of the tapered portion. The JT1 taper is impossible to machine by dead reckoning. Instead, replicate the taper of an existing straight-to-JT1 adapter (see Section 4-34).

With the straight shank of the adapter in your best chuck or collet, set the compound angle against the taper using a sensitive dial indicator. When the needle shows no movement as you traverse the taper (or as close to no movement as you can get), cut an *experimental taper* on soft material before moving to the ground rod—but don't jam the chuck onto your trial taper so hard it can't be budged.

What then? The answer: Make up a couple of U-shaped wedges to slip over the spindle, nose to nose; then tap them together just like shimming a door frame (Figure 10-3).

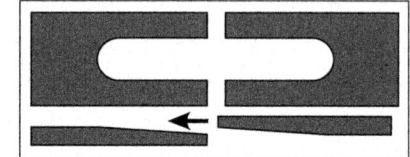

FIGURE 10-3 U-shaped wedges. Make these in the shop from standard plastic wedges used in construction.

With your trial taper still in the lathe, make sure of the fit by coating the taper with marking fluid or a fiber-tip pen. Let it dry; then press the chuck onto it with a rotating action. Separate the two; then inspect both for signs of inconsistent rubbing. Before cutting the real thing, be sure your knife tool is capable of a surface finish ready to go with nothing more than minor dressing with a diamond lap.

10-6 BACK STAY AND LINK COMPONENTS

The back stay and main link were 1-1/4" x 1/2" steel (1018). The connector, from the same material, is 1/2" thick x 0.6" wide. Center to center on all frame components is 2.25". The drill rod was used for the three pins, 0.25" for the bottom two and 0.188" for the one through the arbor. The center of each pin was flatted 50 thousandths or so to locate the #4-40 set screws without marring.

The back stay is bored for a tight press fit on the sleeve, either as it came from the supplier (probably 0.82" diameter) or necked-down to provide a shoulder.

In either case the back stay is dry-fitted to the sleeve, which is then through-drilled undersize for a 1/8" drill rod pin. After cleanup, the frame member is "Loctited" (permanent grade) and pinned to the sleeve. If you are nervous about hole-to-hole alignment, you might want to drill for the pin after assembling with Loctite. The oval cutout in the main link is 0.5" clearance for the rod and measures 1.1" overall (you need this length for full articulation).

> **Important:** I have seen other versions of this type of linkage in which the main link was cut top to bottom, tuning-fork style. I don't like this because cold-drawn steel, when stress-relieved, springs open like a flower, 100% predictably.

The handle is a 3/8" steel rod, shouldered and threaded 5/16-24 for insertion into the main link. In my case, the other end of the rod is capped with a 1"-diameter ball I had on hand. (McMaster Carr refers to these balls as "plastic threaded ball knobs.") In the photo, Figure 10-1, the rod was about 6" overall, but has since been cut down to 4-1/2" for better control of drill pressure. (A smaller ball knob might be better, too.)

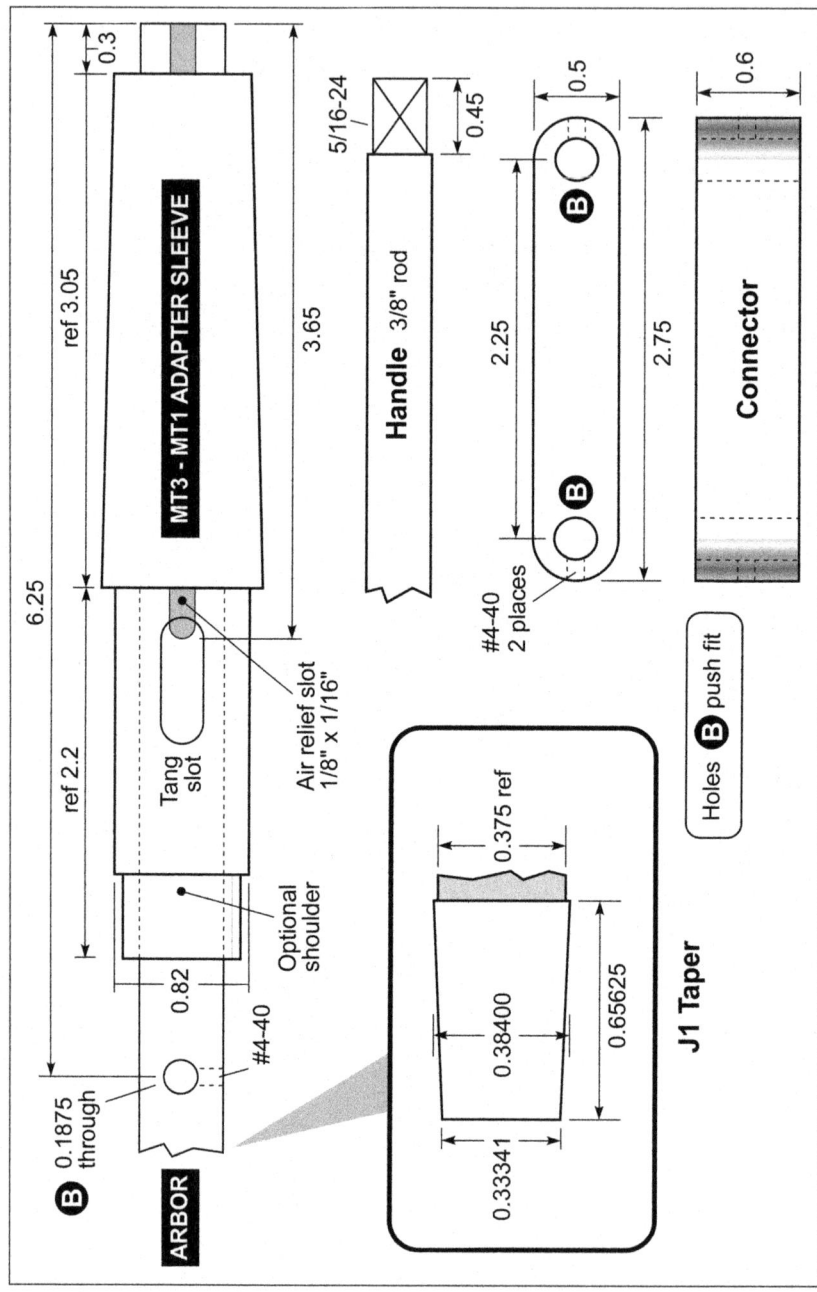

FIGURE 10-4 (1) Suggested dimensions for MT3 taper drill press.

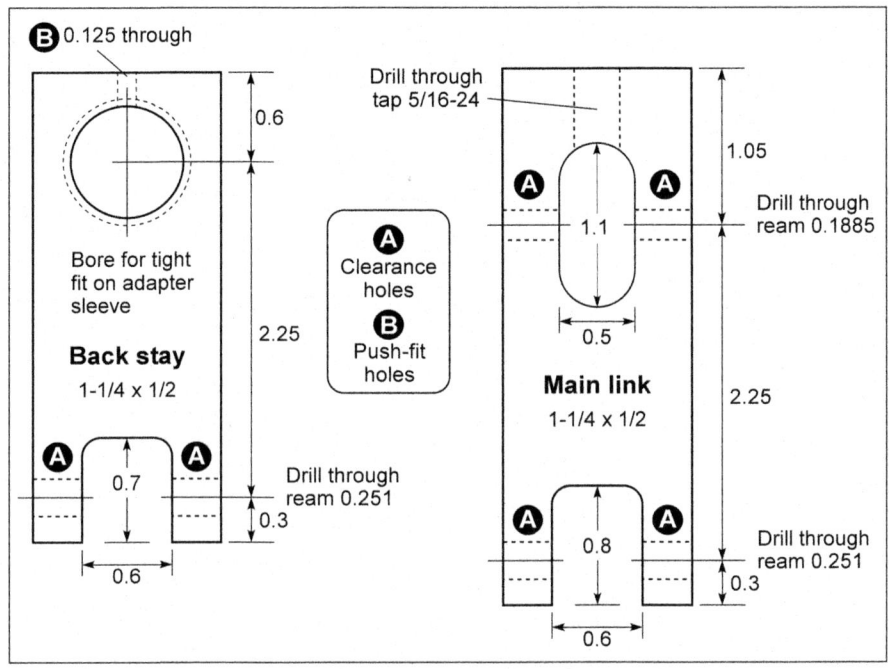

FIGURE 10-4 (2) Suggested dimensions for MT3 taper drill press (*continued*).

CHAPTER 11

A Toolholder for Easy Screw Cutting

CONTENTS AT A GLANCE

11-1 BACKGROUND AND HISTORY

If you do a lot of single-point screw cutting, mostly external threads, you find yourself performing a tiresome ritual: (1) Make a cutting pass, (2) disengage, (3) retract the cross-slide, (4) return the saddle, (5) add a few thousandths infeed on the compound, (6) reset the cross-slide, and (7) reengage. And it doesn't take long before you begin to wish for something less monotonous.

Some of the fancier toolroom lathes automate the process to some degree by disengaging at a specific point, even retracting the cutting tool for the return pass. However, don't look for such features on today's low-priced imports.

What you can do instead is make a retracting toolholder that goes a long way in lightening the load. The design described in this chapter is a Series 100 size toolholder with a self-contained retracting lever—in effect a stand-alone special-purpose compound, designed for one angle setting, namely 29°. The one thing this device does is to eliminate the cross-slide retraction and resetting—set the cross-slide once at the start of the process; then forget it.

The toolholder described here was originally left-facing, in line with the traditional compound setting angle recommended for the past 100 years or so. This was referred to in Chapter 5 as Method A. Figure 11-1 shows the original version, in use for 10 years or more.

More recently I was sold on the "less stressful" way of screw cutting, Method B, in which the saddle moves to the right, clearing the workpiece completely instead of stopping—if fortunate—at a runout groove (Method

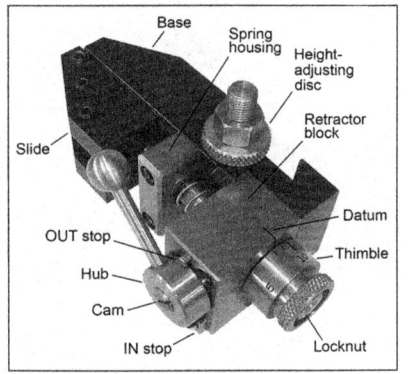

FIGURE 11-1 Left-facing retracting 29° toolholder.

A). The result of that rethink was a new slide that faces in the other direction (Figure 11-2). Everything else in the current toolholder (Figure 11-3) was taken from the original version.

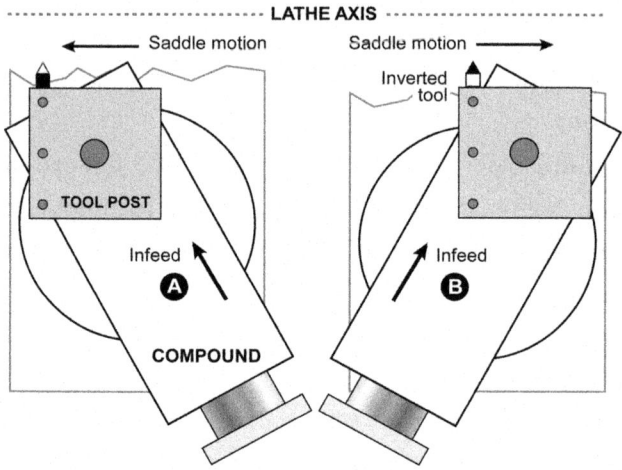

FIGURE 11-2 Traditional (A) and preferred (B) compound settings. In Method B, the screw cutting action is to the right, and the tool is inverted (shown in Figure 11-3).

FIGURE 11-3 Right-facing 29° retracting toolholder. This is used mostly for cutting right-hand external threads, with the cutting tool inverted (indicated by the arrow).

The general assembly of the new toolholder is shown in Figure 11-4. For an exploded view of the components, see Figure 11-5.

A TOOLHOLDER FOR EASY SCREW CUTTING

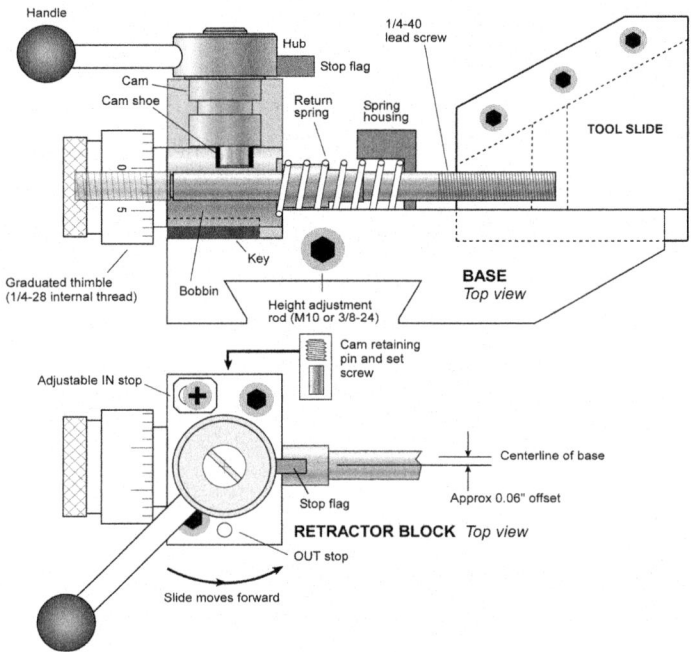

FIGURE 11-4 Retracting toolholder general assembly.

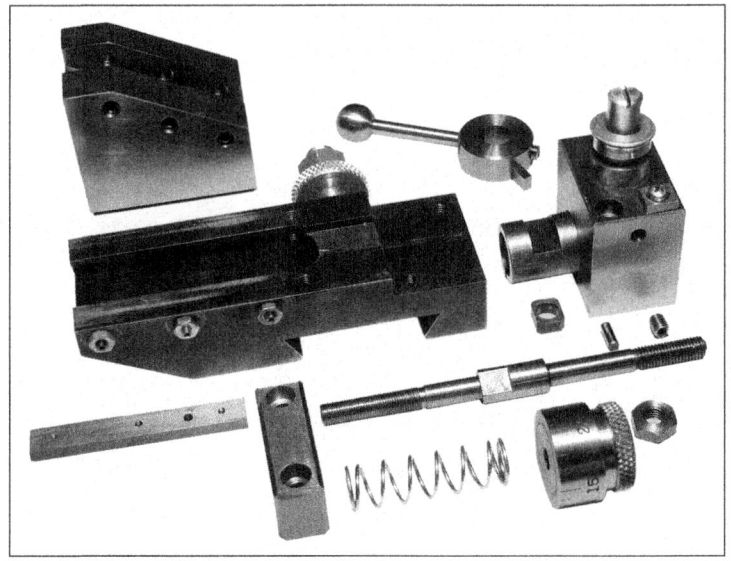

FIGURE 11-5 Retracting toolholder components.

11-2 THINKING OF MAKING THIS TOOLHOLDER?

If you use your lathe for screw cutting only occasionally, say, every other week, it would be hard to justify the effort—other than as an interesting exercise in shop work. On the other hand, if your need for screw cutting is fairly frequent (and you dread the tap dance that goes with it), you may find this special tool worth the effort. You will need access to a small mill, but there are no special tool requirements other than a 3/4" 60° dovetail cutter.

In the following sections of this chapter, you'll find brief how-to notes and drawings. A more detailed description of the original version (left-facing) was published in two issues of *Home Shop Machinist*, July/August and September/October 2017.

11-3 DOVETAILS AND OTHER DETAILS

If you are comfortable milling 60° dovetails, there is nothing here to worry you. In theory, most of the parts in this project can be made as drawn, but there are instances where precise measuring is not quite enough. In some cases, it is better to fine-tune as you go on the basis of "A fits B, B can't interfere with C," and so on. Examples of the fit-as-you-go approach are the dovetail slots in the toolholder base and tool slide, and the width of the retractor block.

Most dimensions in this chapter are given in multiples of 0.001", so 250 equals 0.25".

11-4 A TRIAL RUN CAN AVOID MISTAKES

Like everything else in engineering, the finished product is the best compromise of competing factors. One way to be sure you have the right dimensions—which includes verifying those given here—is to model the entire project in easy-to-work materials such as rigid PVC. This is a low-

cost material that machines beautifully at high spindle speeds, with no
significant cutter wear. One valuable
bonus you get from working in PVC
is the ability to practice machining
procedures with the least possible risk
(Figure 11-6).

Before cutting metal . . .

FIGURE 11-6 Practice machining on PVC.

- Know that this 29° toolholder,
  shown above in Figure 11-1,
  was intended to be used with
  a solid base on the cross-slide (shown in Chapter 3, Figure
  3-1), but there is no reason it could not be used with a conventional compound.

- Take care to allow *exactly* for the offset from the centerline
  caused by the gib, which (as built) is on the underside of the
  base—*but it does not have to be.* The offset is nominally about
  0.06", but that will vary with the gib material thickness.

- Be sure that the dimensions given allow sufficient vertical
  adjustment for the various sizes of tool shanks (1/4", 3/8",
  etc.), both face-up and face-down. Note that the dovetail slot
  is symmetrical in the toolholder base. The gib strip offsets the
  tool slide by about 0.06".

- Watch out especially for the *UP* travel limit, the point where
  the knurled height-setting disk is as far down as it will go.
  You may find that the face-down threading tool, for instance,
  sits too low.

- Like other "mostly milling" projects, this will go more
  smoothly if the two main workpieces are first prepped as
  rectangular blocks, accurately squared up to their exact outside dimensions before you do anything else.

- As built, the three gib screws are on the underside of the tool slide. More conveniently, they could be on the upper surface.

11-5 A WORD ON DOVETAIL CUTTERS

A dovetail cutter is not like an ordinary end mill. Axial cutting forces tend to pull it out of the spindle. One fix for this is to use an *end mill holder* instead of a collet, Figure 11-6. Wiggle the cutter down firmly against the holder's set screw before fully tightening. It is assumed here that the cutter is a standard ANSI 3/4" 60° cutter (Figure 11-7). The angled flutes measure only 5/16" along the axis, so the 3/8"-deep slot in the base will have overhanging lips that must be removed.

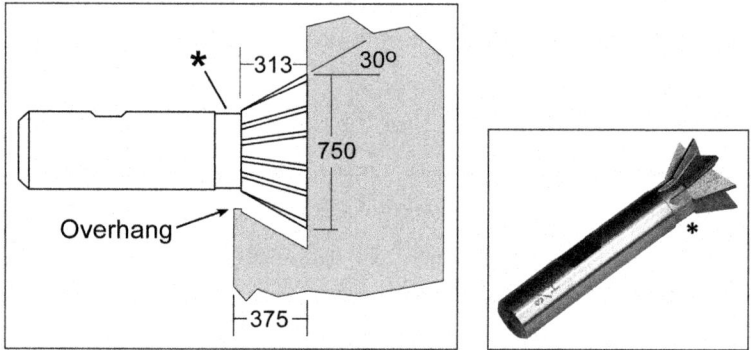

FIGURE 11-7 Typical dovetail cutter. Beware the parallel cutting portion indicated by the asterisks—this can mislead you into overcutting a dovetailed cavity.

11-6 GIB

Make this first, because you will need it to test-fit the tool slide in the toolholder base. Start with mild steel sheet 0.3" x 2-1/2", ideally about 3/32" (0.094") thick.

This translates to a width directly across the dovetail of 0.094/cos 30°, approximately 0.11". A thicker gib will require more material to be removed from the slide or the base.

Figure 11-8 shows one way to machine the gib. The gib, overlong initially, is clamped at both ends to a scrap of 1/2"-thick aluminum by two screws. Using the 60° dovetail cutter, machine about 1/8" off the top surface of the aluminum, leaving a narrow ridge along the back edge. This will locate the gib accurately and also minimize upward bowing. Both the 275 dimension and flatness of the gib are important. Also note that the outer edge of the gib must lie below the surface of the dovetail slot in the base.

Use this setup to mill the strip to thickness, as well as milling the 60° edges.

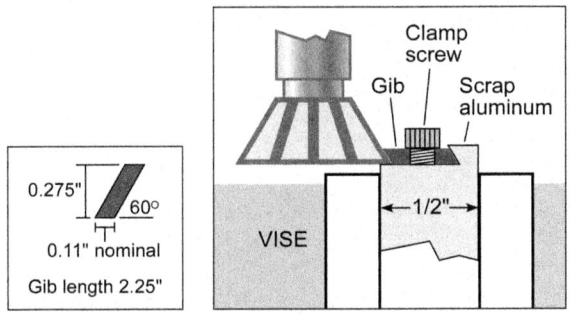

FIGURE 11-8 Gib machining (gib dimensions inset).

11-7 MACHINING THE TOOL SLIDE

You may want to practice machining this on PVC. (The same applies to the base, shown in Figure 11-10 later in the chapter.) Aside from anything else, a PVC tool slide will be useful as a clamp to hold the gib in place when drilling the base for the gib locating pin.

My tool slide (Figure 11-9) is a heavy chunk of steel, unnecessarily so, it seems to me now. If I were making another, it would 1-3/4" long (even 1-1/2") instead of 2". For the version as drawn, start with a squared-up block, 2" x 2" x 1-1/4".

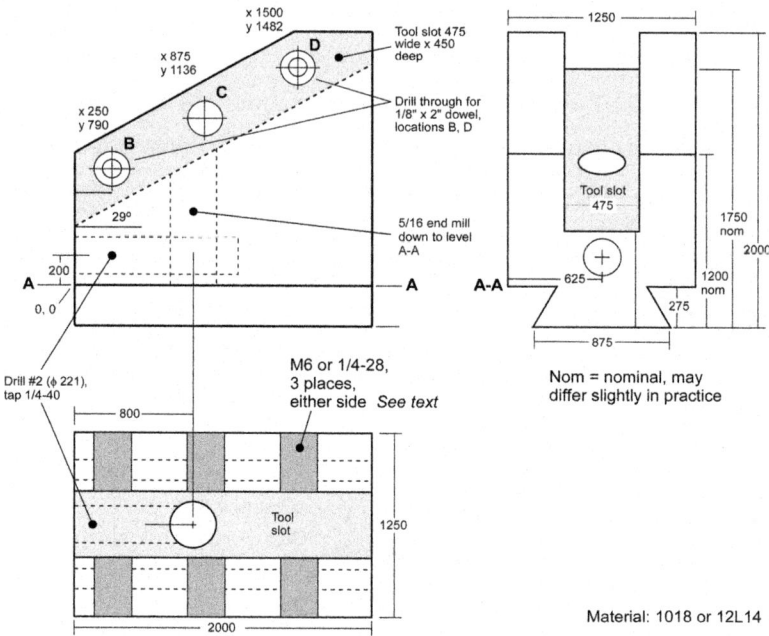

FIGURE 11-9 Tool slide.

Drill 1/8" through holes at B, C, and D. Holes B and D are for dowel pins that will later (final operations) rest on the mill vise jaws to set the block at 29° to cut the sloping top surface, followed by the tool slot.

The vertical 5/16" hole 800 from the front surface defines the inner end of the 1/4-40 lead-screw nut and also provides access for oiling. Drill 5/16"; then (optional) end-mill down to A-A, just below the threaded hole.

To save wear on the dovetail cutter, remove the surplus material to a depth of, say, 260 with a regular end mill. Don't climb mill when cutting the dovetails. Start dovetail-cutting *at the depth cut by the end mill.* Don't try to remove a lot of material each pass. Climb milling does not work well with dovetail cutters.

The following procedure for the dovetails may help:

1. Referring to the Y axis dial or DRO, locate the centerline of the workpiece.

2. Bring the workpiece forward just clear of the cutter.

3. With the spindle running, move the workpiece back to engage the cutter by a few thousandths.

4. Note the Y offset from the centerline, and then make the first (right-going) cutting pass.

5. Move the workpiece back to exactly the same Y offset in the other direction, and then make the second (left-going) cutting pass.

6. Repeat steps 4 and 5, incrementing a few thousandths each time, until the widest dimension of the slide is a shade over 900.

7. Lower the cutter to 275, and then make the additional passes in each direction to finish the job. Make sure that the sliding surfaces are smooth and meet in a sharp corner—no steps.

Take *special care* when drilling the 0.221" hole to be tapped 1/4-40. Misalignment will tilt the lead screw off-axis, which will cause it to bind.

11-8 MACHINING THE BASE—QCTP DOVETAIL

Machining the base (Figure 11-10) calls for a large amount of material removal that may cause bowing on the long dimension. This may need to be flattened before machining the slide ways.

1. Drill 400 deep for the post; then flatten the well with an end mill. Tap 3/8"-24 or M10 with a bottoming tap. (I used 3/8"-24, shown in Figure 11-11, because both tap and threaded rod were on hand.)

2. Establish the target dimension of the slot by measuring your other QCTP holders using 1/4" ground rods/dowel pins (Figure 11-11), not by measuring across the lips.

3. Using an end mill, rough out the slot for the QCTP post 360 deep. Make the first passes on the 1350 centerline; then open up to 1200 wide, cutting equal amounts each side of the centerline.

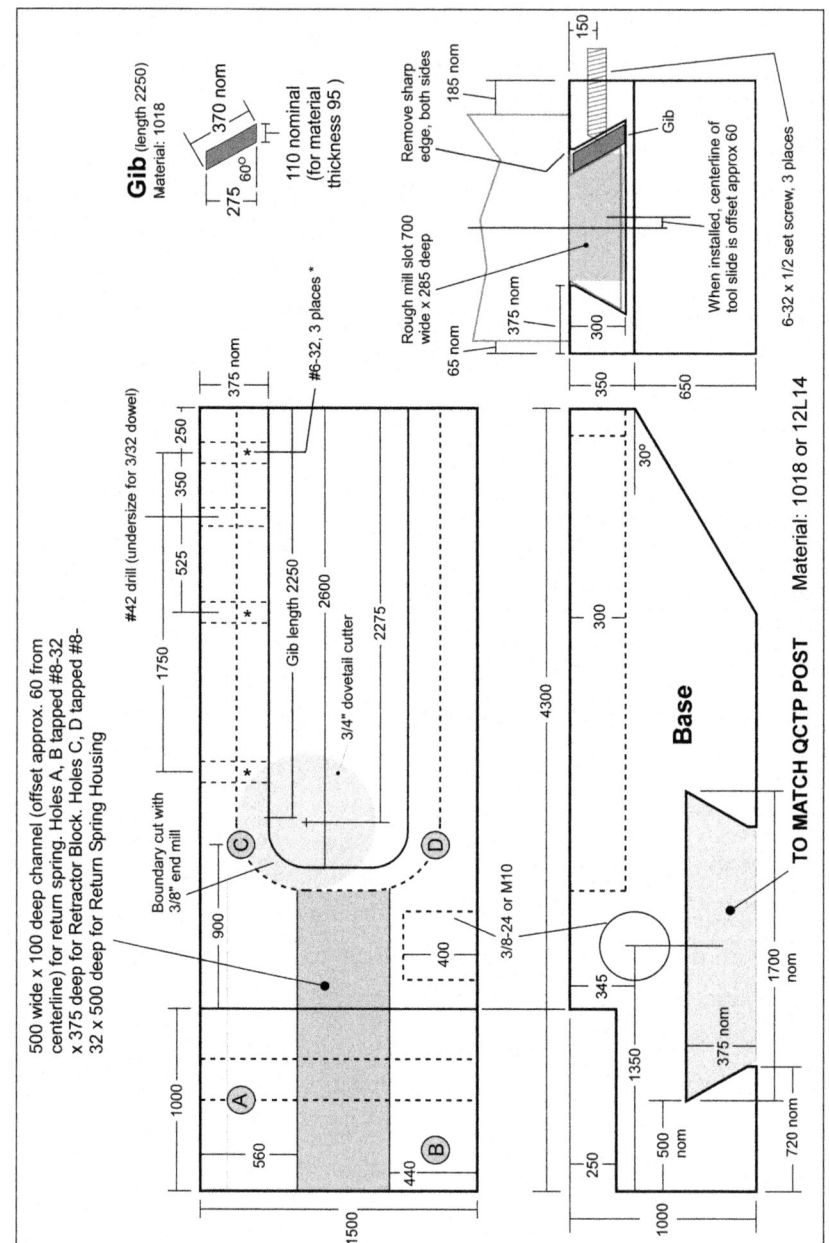

Gib (length 2250)
Material: 1018

370 nom
275
60°

110 nominal
(for material
thickness 95)

Remove sharp
edge, both sides

Gib

When installed, centerline of
tool slide is offset approx 60

6-32 x 1/2 set screw, 3 places

Rough mill slot 700
wide x 285 deep

185 nom

375 nom

300

150

65 nom

350

650

500 wide x 100 deep channel (offset approx. 60 from
centerline) for return spring. Holes A, B tapped #8-32
x 375 deep for Retractor Block. Holes C, D tapped #8-
32 x 500 deep for Return Spring Housing

#42 drill (undersize for 3/32 dowel)

#6-32, 3 places *

375 nom

250

350

525

1750

Gib length 2250

2600

2275

3/4" dovetail cutter

Boundary cut with
3/8" end mill

900

400

3/8-24 or M10

1000

560

440

1500

(A)

(B)

(C)

(D)

4300

300

30°

Base

345

1700
nom

1350

375 nom

500
nom

720 nom

250

1000

TO MATCH QCTP POST   Material: 1018 or 12L14

FIGURE 11-10   Base.

| 260 |

4. With the 60° dovetail cutter just skimming the bottom of the end-milled slot, take light cuts first on one side and then the other for symmetry about the centerline. Measure repeatedly as you go with the 1/4" rods.

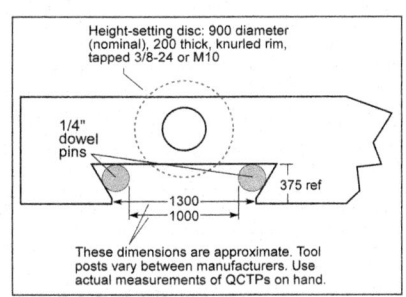

Height-setting disc: 900 diameter (nominal), 200 thick, knurled rim, tapped 3/8-24 or M10

1/4" dowel pins

1300
1000

375 ref

These dimensions are approximate. Tool posts vary between manufacturers. Use actual measurements of QCTPs on hand.

FIGURE 11-11   QCTP dovetail.

5. When you are approaching the target dimension, but with still quite a way to go, lower the cutter to the 0.375" final depth; then skim the bottom and sides.

6. When you are close to the target dimension—within a few thousandths—replace the dovetail cutter with an end mill, and skim off the overhanging lips, as shown above in Figure 11-7. Take off just enough to ensure that no lip will remain when the 60° sides are to size.

7. Reinstall the dovetail cutter. Take very light cuts, a couple of thousandths or so each side, trying to insert the QCTP post after each pass. Scrape/file/clean before inserting, allowing time for cooling. Be careful: Very little off the sides is all it takes to go from perfect to sloppy. The job is done when the post slides smoothly into the slot, locking firmly when the handle is cranked.

8. Check for flatness along the length of the base. Skim both sides if necessary.

11-9 MACHINING THE BASE—TOOL SLIDE DOVETAIL

To machine the base, follow these steps:

1. Rough out the slot with an end mill to 275 deep by 650 wide.

2. At the U-turn end of the slot, cut the boundary at the full 300 depth with a 3/8" end mill, width to blend nicely with the fin-

ished edges of the slot. This gives the dovetail cutter less work to do, allowing it to machine more freely, full width, to 2275 from the open end (a little beyond the length of the gib strip).

3. Using a procedure similar to the steps in Section 11-8, mill the dovetails symmetrically relative to the centerline at 275 depth.

4. When the slot is almost wide enough to accept the tool slide, lower the cutter to 0.3".

5. Preserving the same symmetry relative to the centerline, increase the slot width to the point where the tool slide with its gib strip can just be inserted. Machining burrs and debris can mislead you, so scrape and file before testing with the tool slide.

6. Take very small finishing cuts on each side until a 0.01" feeler gauge can be inserted into the gap between tool slide and gib strip.

7. Drill and tap #6-32 for the three gib screws.

8. Clean up the dovetail slot as necessary for smooth movement of the tool slide.

9. Using an end mill or file, blunt the sharp outer edges of the dovetail. This will ensure good seating of the slide.

10. Left-right motion of the gib is prevented by a drill-rod pin not larger than 3/32" diameter. Drill the base undersize, say #42, for a push fit.

11. If you have a PVC "practice" tool slide, install it with the gib strip; then tighten the gib screws. Drill through the gib strip for the 3/32" pin.

12. Hone the tip of the 3/32" pin to fit easily in the hole just drilled in the gib strip.

13. Press the pin into position, taking care that it does not protrude beyond the gib's working surface. Use Loctite on final assembly if necessary.

14. Oval-point set screws work best for the gib. *Optional step:* For better seating, dimple the gib a few thousandths at each screw location using a #6-32 tap-size drill. Locknuts are required for the screws.

11-10 RETRACTOR BLOCK

Because it's easier to fine-tune a turned diameter than a bored hole, make the retractor block (Figure 11-12) before the bobbin and cam. Start with a block 1.52" x 1.4" x 1.0". (The oversize 1.52" dimension allows for possible misalignment between the slide and base centerlines—the excess is removed after assembly.)

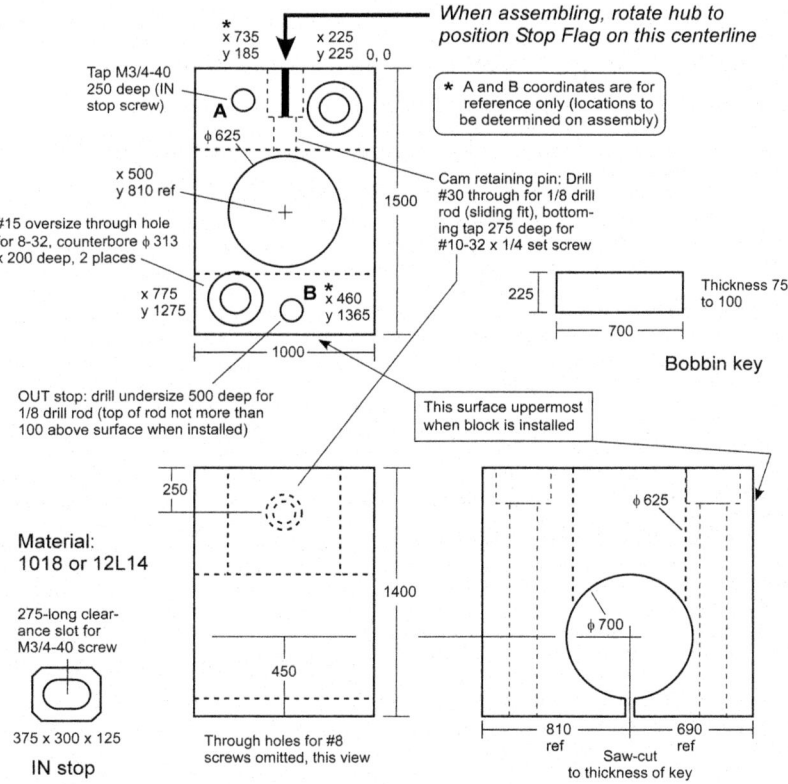

FIGURE 11-12 Retractor block, bobbin key, and IN stop.

No allowance is made for misalignment in the vertical axis, so take care with the 450 centerline for the bobbin bore (diameter 700). The reference dimension 810 in the horizontal axis assumes a nominal 60 centerline offset (your result may vary depending on how you machined the dovetails). Take care also with the hole location for the cam retaining pin. This will make machining of the cam more predictable.

The bobbin key was cut from a scrap of steel about 0.075" thick, and the retractor block was saw-cut for a tight fit. The retractor block is attached to the base by two #8-32 x 1-1/2" socket head screws. The outer face of the block needs a datum line for the graduated thimble (not shown in Figure 11-12). A deep scribed line will do the job (I made a 0.02"-wide saw cut, at 45°, to mark both upper and outer surfaces).

No allowance is made for misalignment in the vertical axis, so take care with the 450 centerline for the bobbin bore (diameter 700). The reference dimension 810 in the horizontal axis assumes a nominal 60 centerline offset (your result may vary, depending on how you machined the dovetails). Take care also with the hole location for the cam retaining pin. This will make machining of the cam more predictable.

11-11 BOBBIN AND SHOE

Turn the bobbin for a close sliding fit in the retractor block. Saw-cut for a sliding fit on the key (Figure 11-13). The shoe needs to slide freely in the milled slot in the bobbin, but with *negligible side-to-side play.*

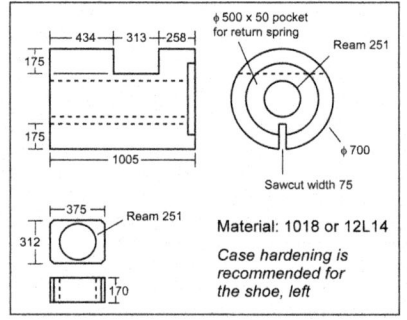

11-12 SPRING HOUSING

FIGURE 11-13 Bobbin and shoe.

A word on the spring itself: Mine has an uncompressed length of 1.5", an OD of 0.48", and eight turns of 0.046"-diameter wire, rate about 12 lbs/

inch. This works well enough, but a slightly higher rate would be desirable. The housing is attached to the base by two #8-32 x 3/4" socket head screws (Figure 11-14).

11-13 HUB AND HANDLE

The recommended five-step procedure for the hub and handle (Figure 11-15) follows:

1. Make the stop flag first.
2. Machine the slot on the mill—go for a really tight fit—before turning the hub on the lathe. Start with 1" stock long enough to be held in a pair of Vee blocks; then mill or saw-cut the slot.

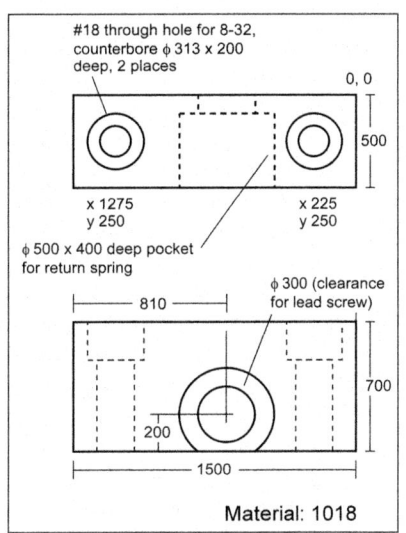

FIGURE 11-14 Spring housing.

3. Drill, ream, and part off the hub. Install the flag with Loctite if necessary. Angular locations of the 3/16" and #10-32 holes are not supercritical, so you may not need a dividing head on the mill.
4. For finish machining, set the hub/flag assembly on a 3/8" stub axle, securing it with a #10-32 set screw. Skim the underside, including the flag; also clean up and chamfer the parting surface.
5. The handle knob can be an off-the-shelf item or DIY using whatever material you prefer. Mine was turned from a scrap of 1/2" 12L14 drilled undersize, force-fit on a 3/16" drill rod.

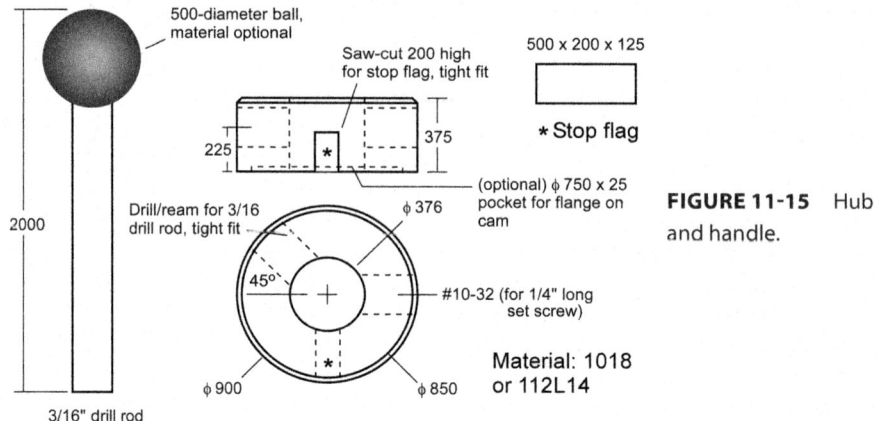

500-diameter ball, material optional

Saw-cut 200 high for stop flag, tight fit

500 x 200 x 125

∗ Stop flag

375

225

∗

2000

Drill/ream for 3/16 drill rod, tight fit

φ 376

(optional) φ 750 x 25 pocket for flange on cam

#10-32 (for 1/4" long set screw)

45°

Material: 1018 or 112L14

φ 900

φ 850

3/16" drill rod

FIGURE 11-15 Hub and handle.

11-14 CAM

As drawn in Figure 11-16, the throw of the cam is 0.85" x 2 = 0.17". This translates to 0.15" along the cross-slide axis (0.15" = 0.17" x cos 29°), way more than needed to clear any 60° threads in the model shop (a 6-TPI thread, for instance, is less than 0.1" deep). For better leverage, consider going with a smaller cam offset. Pay special attention to the 128 groove intended for the cam retaining pin.

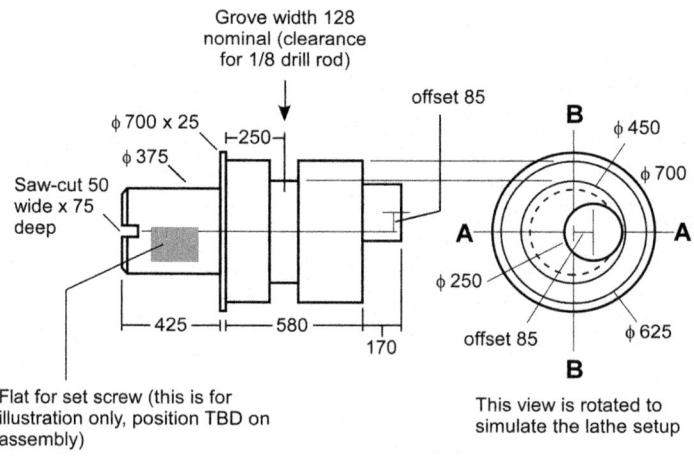

Grove width 128 nominal (clearance for 1/8 drill rod)

φ 700 x 25

offset 85

B

φ 450

250

φ 375

φ 700

Saw-cut 50 wide x 75 deep

A

A

φ 250

425

580

170

offset 85

φ 625

B

Flat for set screw (this is for illustration only, position TBD on assembly)

This view is rotated to simulate the lathe setup

FIGURE 11-16 Cam.

If this is even a little off, the pin may not be insertable. Or worse, it may jam so tightly in the groove that it cannot be removed—fatal problem! A *better idea:* instead of a drill rod pin, make a custom #10-32 set screw with a screwdriver slot and a 1/8"-diameter cylindrical end.

The following assumes a cam offset of 85, as drawn in Figure 11-16, but you may decide on a smaller offset.

1. Starting with 3/4" rod, accurately centered in a 4-jaw chuck, finish-machine a 1300-long x 700-diameter section; then reduce the outer end to an 800-long x 625-diameter section for an *exact fit* in the cam bore of the retractor block.

2. Machine the 450-diameter x 128-wide groove with its center-line 250 from the shoulder.

3. Using a miniature level, set the jaws of the stationary 4 jaw symmetrically horizontal/vertical.

4. Set a dial indicator against the 625-diameter section, on the centerline A-A.

5. Adjust the cross-slide to preload the indicator by at least 100; then note the indicator reading.

6. Offset the workpiece in the 4 jaw by 85 along A-A.

7. Offsetting on A-A may have slightly displaced the workpiece on the B-B axis. With B-B horizontal, check and adjust to give the same indicator reading on both sides. Recheck A-A for the desired 85 offset.

8. Machine the offset portion 170 x 250 for a close fit in the shoe (preferably case-hardened).

9. Part off at 1300 from the end. Reinstall the workpiece, flipped, with minimum TIR before machining the 375-diameter section. Aim for a push fit in the handle hub.

10. Finish the 375-diameter section 425 long from the 700-diameter flange; then saw-cut a screwdriver slot at the end.

(This will allow the crankshaft to be rotated into position before installing the handle.)

Optionally, case-harden the cam, but not until the toolholder has been fully tested.

11-15 THIMBLE

Division lines on the thimble are a personal choice; mine are 125 long, with every fifth line 200; the lines were 10 thousandths deep, cut with a 40° included-angle Vee tool. If you don't have a dividing head on the lathe, it may be possible to get by with a hand-drawn cardboard protractor attached to the lathe spindle. Fanatical precision is not necessary.

As drawn (Figure 11-17), the thimble is not zero-settable. I have two other versions that are, but I have never used that feature. The thimble is adjusted to minimize end float of the bobbin, and then is secured by a recessed locknut. (Be sure the recess is large enough for a 7/16" A/F box wrench.) The combination of a wrench on the locknut and a 7-mm wrench on the flats makes for a really solid assembly.

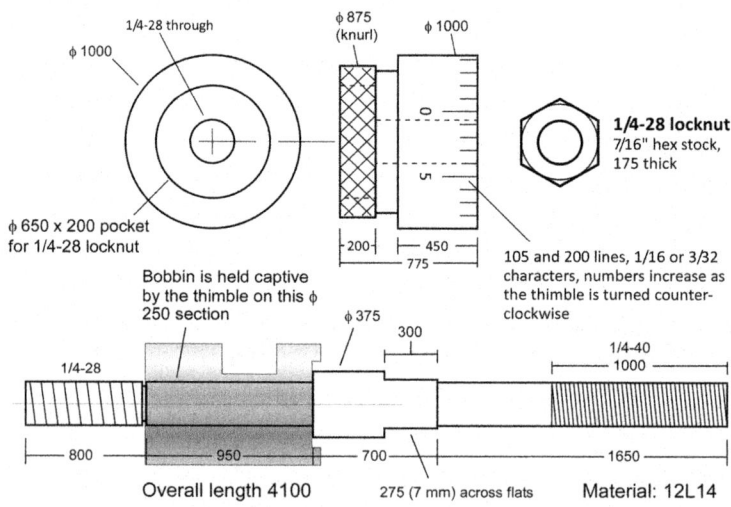

FIGURE 11-17 Lead screw and thimble.

11-16 LEAD SCREW

The 1/4-40 thread was chosen because it allows well-separated 0.001" divisions on a small 25-division thimble. If this is a one-time use, you can get by with a non-HSS tap. You may not need a die to go with it—screw-cut on the lathe instead. If you do use a die to clean up the thread, *open it up to the max* before the first pass (otherwise, you will have too loose a fit in the tool slide).

The thimble is internally threaded 1/4-28 and is secured to the lead screw with a locknut.

Concentricity of the 1/4-40 section and the 250-diameter section (inside the bobbin) is important for smooth rotation of the lead screw.

Here is a suggested machining procedure for the lead screw:

1. Cut a 1/2"-diameter 12L14 rod to the 4" (approximate) finished length, plus enough surplus to allow the knob end to be held in a 3-jaw chuck, with enough room to turn the 1/4-28 and 250-diameter sections in the same setup (Appendix, Section A-7), using both RH and LH knife tools.

2. Center-drill the end of the rod to be threaded 1/4-40.

3. Rough-machine 0.02" oversize on diameter at both ends of the rod with the knob end held in the chuck and the other end on a live center.

4. Cut flats on the 375-diameter section on the mill, or file them in place.

5. Set up the rod as before on the lathe (3 jaw + live center); then finish-machine all diameters. Take special care with the 250-neck diameter.

6. Screw-cut the 1/4-40 thread. With the compound set at 29° or 30°, the infeed for full depth is only 15 thousandths or so. Stop at, say, 0.012"; then brush the thread clean before test-fitting in the tool slide.

7. Make repeat passes in half-thousandth increments, testing the fit each time. To minimize backlash, the objective is a snug fit in the tool slide. Be sure the nut and screw are both clean and lightly lubed before testing. With this tiny depth of thread, minute particles are significant. Bear in mind that the fit eases quickly after only a couple of passes.

8. Screw-cut the 1/4-28 thread; then part off. Finish the 1/4-28 thread with a die if necessary.

11-17 STAMPING THE NUMBERS

This is the point where most how-to guides become vague, or the writer tiptoes away—like the math professor who, while heading for the door, announces that solving the next problem is up to the interested student (*translation:* Lots of luck, kids!).

Marking metal neatly is difficult to do in projects like this where production methods (silk screening, etching, etc.) don't apply. There are two alternatives:

1. Do nothing, counting divisions as you go.

2. Stamp the numbers, this being what most model engineers try at least once before starting over. I have never been fully satisfied with my stamping, but at least it's functional.

Stamping is unpredictable because:

- From one stamp to another, the number position can vary up and down.
- The workpiece has to be rigidly supported, carefully positioned, and then locked for each stamping.
- For consistent depth of impression, the hammer blow must be tuned to the number. *Example: 1* takes a lighter blow than *8*.
- To stamp a curved surface, the centerline of the stamp must be precisely on a radial of the workpiece.

All of this adds up to a process that is next door to uncontrollable, unless you take the time to build a jig. I used 3/32" numbers, but 1/16" would have been better for this project. Also, you could use aluminum instead of steel for the thimble. If you decide to number the thimble, be sure the numbers increase as the slide *moves forward*—this is easy to get wrong.

11-18 ASSEMBLING THE TOOLHOLDER

After deburring and cleaning, do a dry-run assembly as follows to trim oversize items such as the retractor block. (When this is done, disassemble, clean, and oil everything in sight, and then reassemble.)

1. Start by installing the gib strip and slide in the base, gib screws loose.
2. Assemble the bobbin and thimble on the lead screw; tighten the thimble for minimum end play of the bobbin, and then secure it with a locknut.
3. Install the bobbin key in the retractor block.
4. Leaving the cam, spring, and spring housing aside for the moment, run the bobbin/lead-screw assembly into the retractor block; then thread the lead screw into the slide as far as it will go.
5. Fully tighten the gib screws.
6. With the retractor block "floating" (screws loose), allow the bobbin to self-align on the lead screw. When you are satisfied with the action of the lead screw—no notchiness—tighten the two block screws very firmly. If you suspect that the bobbin is sitting too low or high, correct by shimming or by removing a few thousandths from the underside of the block.
7. If necessary, trim the retractor block overhang (1500 dimension) relative to the base.
8. Remove the finished retractor block and reinstall the bobbin/lead-screw assembly, this time with the spring and spring housing in place.

9. Install the cam shoe, cam, and cam retaining pin.

10. Check the IN/OUT motion of the assembly by turning the cam with a screwdriver.

11. Reinstall the gib strip and slide; tighten the gib screws so that the slide moves with some resistance when pushed with the fingers.

12. Run the lead screw fully into the slide; fasten down the retractor block and spring housing.

13. Turn the cam with a screwdriver to bring the slide fully forward to top dead center (TDC).

14. Test a few times for TDC with a feeler gauge between the thimble and retractor block (the gap should be in the region of 0.015", but the exact amount is not important); at this point, the slide should feel "solid," with no tendency to move in either direction.

15. When satisfied that the cam is truly at TDC, install the hub/handle with the flag on the block centerline, as shown earlier in Figure 11-12.

16. Tighten the hub set screw hard enough to mark the cam axle.

17. Remove the hub/handle. Remove the cam, and set it in Vee blocks in the mill vise with the indentation at TDC. It's important to get this right—otherwise, the flat you are about to mill in the axle may be too far off, radially, to be compensated for by the adjustable IN stop.

18. Mill a flat at least 200 wide x 50 deep centered on the indentation.

19. Reinstall the cam; check that the hub/handle with its set screw on the just-milled flat is where you expected it to be, or if not, within the range of adjustment for TDC provided by the IN stop. (Even if the flag hits the stop well before TDC, there may be no cause for concern if it's no more than 2° or 3° off.)

20. Install the drill-rod OUT stop, making sure it clears the handle.

21. Before final assembly, consider case-hardening at least the shoe axle at the bottom end of the cam, and also the shoe itself. This may not be needed if the assembly will be used only occasionally.

22. Test the final assembly for repeatability with a few dozen IN-OUT cycles, as shown in Figure 11-18 (this is an earlier, left-facing version). The IN reading should not vary detectably. When checking the depth of stroke, bear in mind that a full revolution of the thimble, 0.025" slide motion, will cause a dial indication of only 0.0219" (0.025" x cos 29°).

FIGURE 11-18   Testing the IN/OUT action.

CHAPTER 12

Making a Precision Grinder for Lathe Tools

CONTENTS AT A GLANCE

12-1 KEY POINTS

After years of experimenting with various grinding methods, including some fairly fancy machines, I designed a grinder from first principles (Figure 12-1). It delivers precise, very acceptable results, but it doesn't consume weeks of shop time. If the material is on hand, it can be built in a day or two. Some milling is necessary, but nothing that can't be handled by a small milling machine.

Key points of the grinder are:

- It handles all the basic RH/LH lathe tool grinds, including boring tools, D-bits, etc.
- For a versatile machine, it's about as simple as it can be.

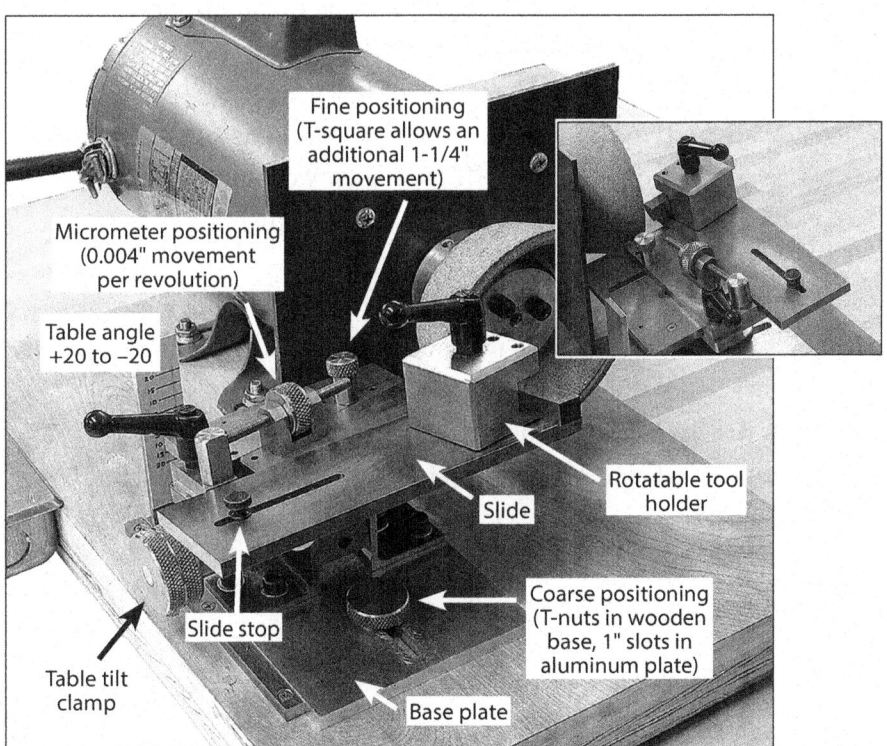

FIGURE 12-1   Precision grinder. Inset: close-up of tool slide assembly, viewed from the rear. The guard was removed for visibility.

- Aside from the motor, the cost can be as low as $75, grinding wheel included, depending on your raw material stock.
- If you want to grind carbide tools, you can add a 100-mm diamond wheel.
- Any standard 1/3-hp grinder motor will do, provided its no-load speed is around 3500 rpm. Figure 12-1 shows a Rockwell motor from an old belt sander, counterclockwise only.
- It is rock solid (1/4"-steel table), allowing you to use as much pressure on the wheel as you need.
- It has three separate adjustments: (1) The base allows *coarse* front-to-back movement of over 1", (2) a T-square on the table provides additional *fine* front-to-back movement of 1-1/4", and (3) a lead screw–controlled wedge on the T-square acts as a *micrometer* adjustment, moving the tool forward in increments of 0.001" or less.
- Aside from the T-square and wedge, there is nothing critical about the dimensions. There is plenty of latitude in the design, allowing the usual home-shop improvisations in using materials on hand.
- Positioning of the tool feed is 100% predictable—and repeatable.
- There are no bearings—the toolholder slides across the table surface, separated by replaceable brass shims (wear strips).
- At the end of a grinding session, it can be taken apart, brushed off, and reassembled in two minutes. This is a real bonus, because grit gets everywhere and is tough to clean up reliably with a shop vac.

12-2 BUILDING THE PRECISION GRINDER

Most projects are conditioned at least partly by what's in the stock pile or scrap pile, this one included. Except for the T-square, micrometer wedge, and related fittings, these notes are suggestions only. Only limited information is given on the motor base and the pivoting support for the table.

(My table was attached with countersunk screws instead of welding, which would have been less work.) Based on a lot of flying time with this grinder, one takeaway is that the locking functions, especially for the table angle, need to be both accessible and rock solid.

Everything in a precision grinder is a compromise of one sort or another—for instance, dealing with the toolholder's various compass bearings affects both the angular range of the table and its fore-aft positioning; another issue is minimizing the vertical shift of the tool face relative to the motor spindle at various table angles and T-square locations (Figure 12-2).

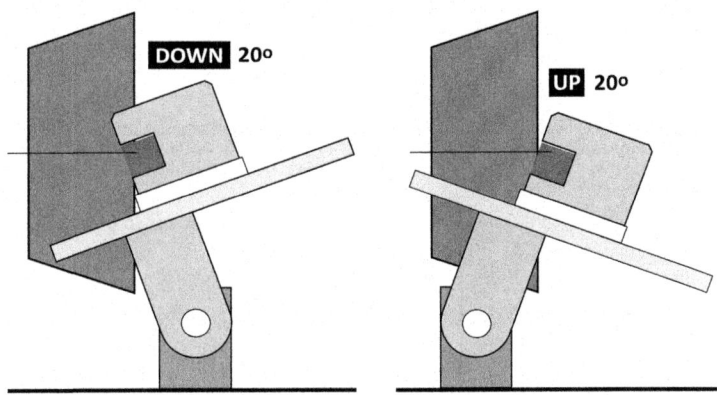

FIGURE 12-2   Table angle versus coarse front-to-back adjustment.

Once you have a suitable motor on hand, measure its centerline height, and build a 2D wood or cardboard mockup of the mounting arrangement, including a dummy table and toolholder, to see how the tool looks at various table angles and fore-aft positions. *One key fact:* In plan view (looking down from above), the right-hand edge of the table sits only about 0.1" to the left of the mounted grinding wheel. This is important, so have the grinding wheel installed on its hub before finalizing the layout.

12-3 MOTOR

Single-phase fractional-hp motors usually come in 1725 and 3450 rpm versions. For grinding, you must go with the *higher speed*, 1/3 hp or more. If you don't have a suitable motor, you can buy an imported bench grinder or a stationary buffer for as little as $35. It's less expensive to discard the bits you don't need than to buy a name-brand motor. In Figure 12-1, a vertical 1/8" PVC panel on standoffs acts as a baffle to protect the motor from the worst of the flying dust.

12-4 GRINDING WHEEL

The grinding wheel (Figure 12-3) is a Type 11 "flared cup," size 4/3 x 1-1/2 x 1-1/4 (larger diameter 4", smaller diameter 3", depth 1-1/2", bore 1-1/4"). For HSS grinding, I use a 60-grit aluminum oxide wheel, Norton 32A60 KVBE, available from McMaster Carr for about $45.

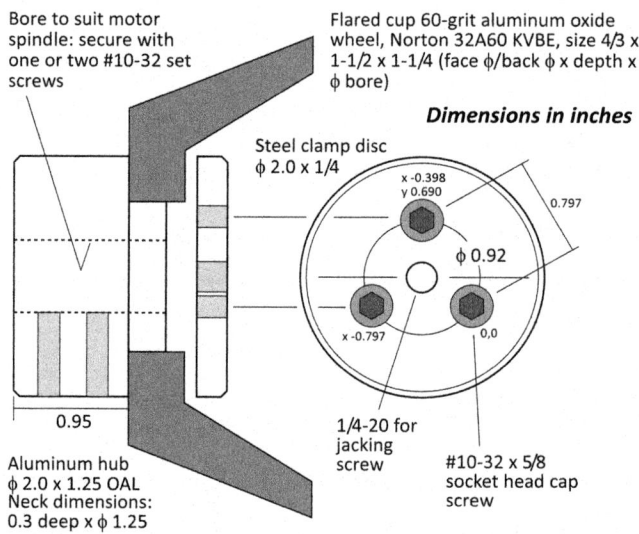

FIGURE 12-3 Grinding wheel and hub.

With the grinding wheel on hand, machine a hub from a 2"-diameter aluminum blank, about 1" long, followed by a matching steel clamp disc (also shown in Figure 12-3). My clamp disc has a tapped hole in the center for a jacking screw to assist in removal of the hub from the motor spindle. (You may never need this, but who knows?) To minimize runout, finish-turn the aluminum blank on a true-running stub axle of the same diameter as the motor shaft. Don't discard the paper washers or labels that came with the wheel—there should be one on the front and one on the back.

If you plan on grinding carbide tools, you will need a diamond-faced wheel (see Section 12-18 at the end of the chapter). Mine is a 100-mm diameter/20-mm bore, $30 or so on eBay. Name brands cost much more.

12-5 BASE PLATE

In my grinder, this is 1/4"-thick aluminum (Figure 12-4). The two long slots allow coarse fore-aft positioning. The tilt scale is pretty basic—a hand-lettered length of plastic strip glued onto a vertical metal strut. The spring-loaded pointer that bears on the scale is attached to the table.

FIGURE 12-4 Base plate. The tilt scale is at the left.

12-6 TABLE

For the table, I used cold-worked 1/4" steel plate (Figures 12-5 and 12-6). My table measures 4" left to right, 4-1/2" front to back. The angle pointer is spring-loaded, ideally by a long, low-rate one as shown in the figures—just enough force to keep the pointer on the tilt scale.

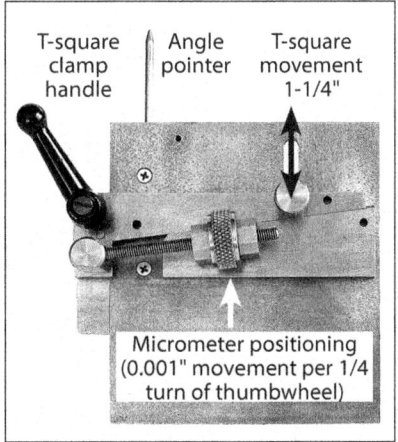

FIGURE 12-5 Table assembly, topside.

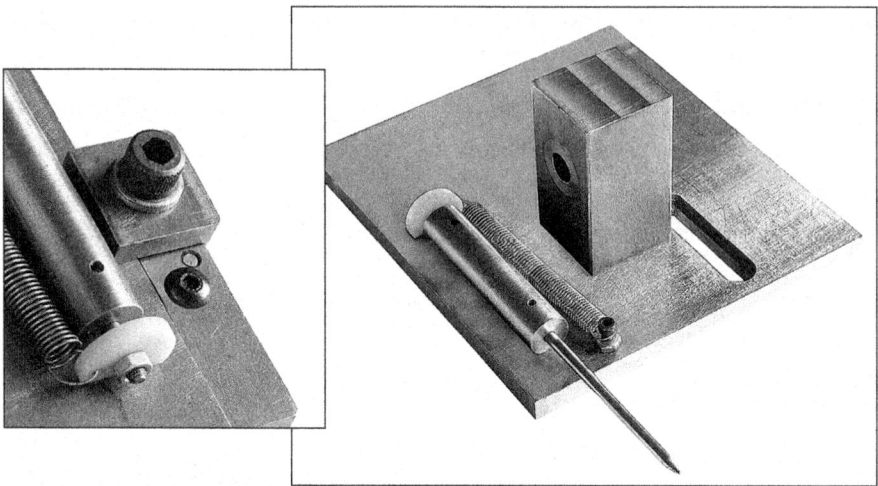

FIGURE 12-6 Table assembly, underside. Inset: close-up of the spring and Delrin washer.

12-7 T-SQUARE, WEDGE, AND GUIDE

This is one part of the project (Figure 12-7) where extra care is called for. Done right, the wedge will be captured solidly under the lip of the T-square and will slide smoothly from side to side, as shown in Figure 12-8. Use flat 1/4" steel plate, checked for consistent thickness. A blank size of 4-1/4" x 1-1/2" will allow you to bandsaw both the T-square and the wedge from one piece, leaving enough margin for milling off sheared or otherwise dinged edges.

Square the blank carefully in the mill, finish-machine its long dimension to exactly 4.25", and then drill both T-square and wedge portions as shown in Figure 12-8, except for the "lead-screw post" hole at bottom left of the T-square. Aim for a push fit for 1/8" dowel pins when drilling the reference holes (filled circles in the drawing); these will be used when machining the ramps. Holes A, B, and C are different: they are used to pin the T-square guide to the T-square, and so they need to be carefully sized for a close fit.

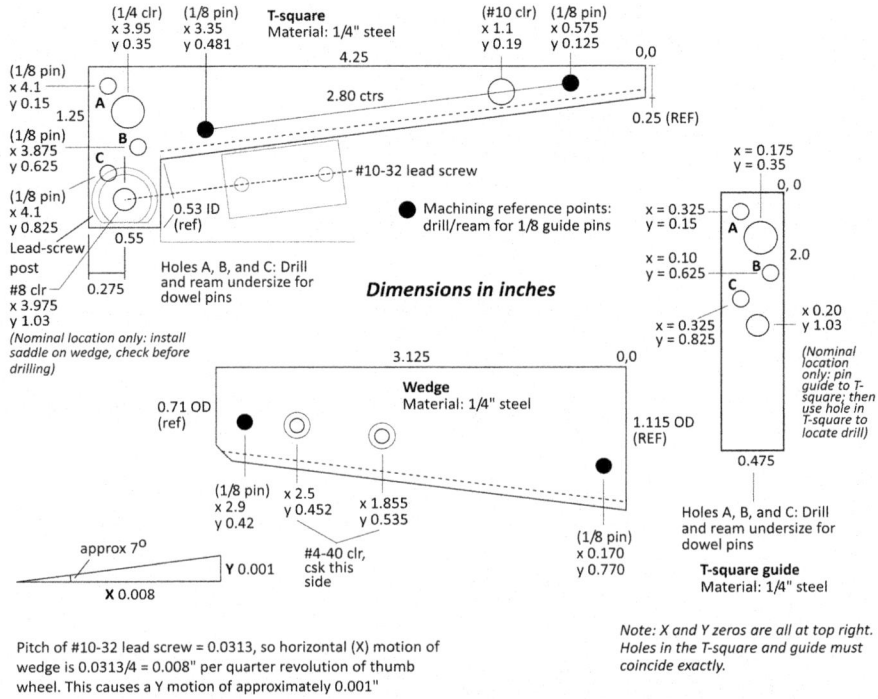

FIGURE 12-7   T-square dimensions.

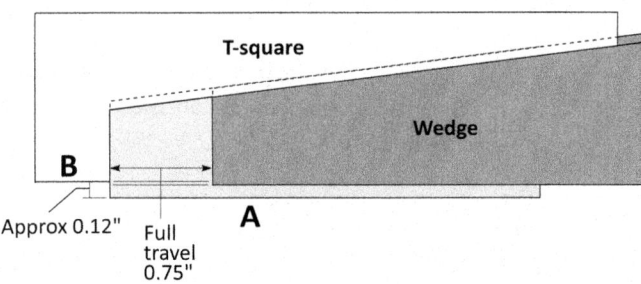

FIGURE 12-8   Caution! Even at full right travel, edge A must be below edge B.

With the drilling completed, separate the parts. This is a good time to think through how the two pieces will be machined. My procedure was (1) install dowel pins in the T-square reference holes; (2) clamp the T-square plate in the mill vise, with the pins resting on the vise jaws; (3) face off the inclined edge with a 3/8" end mill until the narrow end is at the 0.25" reference dimension; (4) undercut the inclined edge, 0.0625" deep x one-half of the material thickness (the lip formed by this undercut has to mate nicely with the corresponding edge on the wedge); and (5) machine the parallel portion of the T-square.

Next, using the same reference pin setup, machine the inclined edge of the wedge to the reference dimension 1.115" at the wide end. Leave the wedge in the vise; then machine its undercut, aiming for a close sliding fit with the T-square (Figures 12-9 and 12-10). Finally, miter the narrow end of the wedge to clear the inside corner of the T-square. Then, with the T-square and wedge together, see if edge A is below B at 3/4" from the home position, as shown in Figure 12-8. If not, it may be necessary to trim the down-going leg of the T-square—or decide to get by with a little less travel.

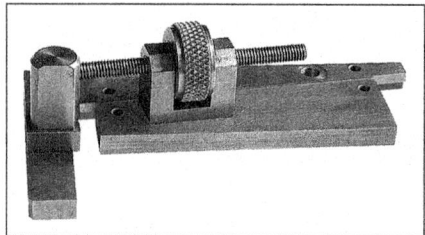

FIGURE 12-9 T-square and wedge assembly.

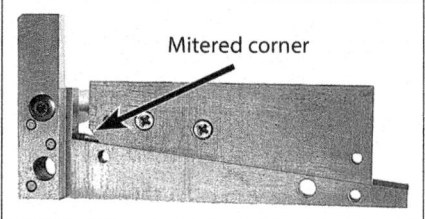

FIGURE 12-10 T-square and wedge assembly underside.

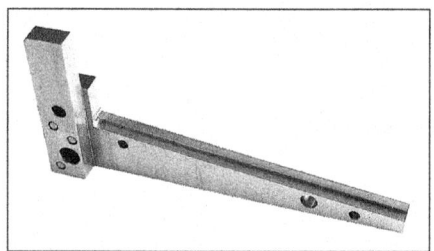

FIGURE 12-11 T-square underside.

The T-square guide is straightforward, but note that holes A, B, and C
are carefully sized for tight-fitting 1/8"
dowel pins; also note that the #8-32
clearance hole for the lead-screw
post is drilled after assembly to the
T-square.

12-8 T-SQUARE FITTINGS

The T-square lock plate (Figures
12-12 and 12-13) is threaded 1/4-20
for a socket head cap screw, with
lock washer, mating with an internal-

FIGURE 12-12 T-square clamp handle and lock plate.

ly-threaded ratchet handle that can be repositioned if it interferes with
other components when clamped. (A second identical handle is used for
the toolholder.) For a close-fitting thumbwheel, make the lead-screw saddle first; then skim the thumbwheel to finish with minimal backlash.

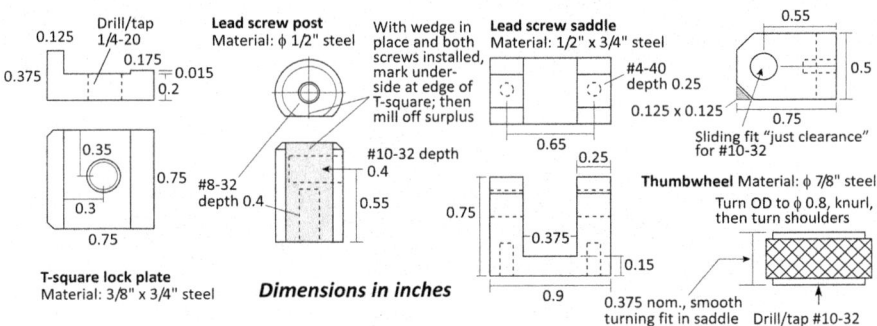

FIGURE 12-13 T-square components.

12-9 ASSEMBLING THE T-SQUARE AND WEDGE

Take care with this. Install the lead-screw saddle on the wedge. Cut a 2-3/4"
length of #10-32 threaded rod; chamfer both ends; then make sure it's

straight by rolling it on a flat surface. The tricky part is locating the center for the lead-screw post on the T-square: Done right, the lead screw will be parallel with the T-square ramp, and the wedge will stay firmly in contact throughout its range of travel.

There are too many variables to do this by dead reckoning, so start by scribing a line down the short edge of the Tee at 0.275" from the left edge, Figure 12-7. Make a "transfer block" as shown in Figure 12-14, with a scribed line (REF) exactly below the threaded hole. Run the straight #10-32 threaded rod through the saddle and transfer block; then mark where the line on the transfer block intersects with the the scribed

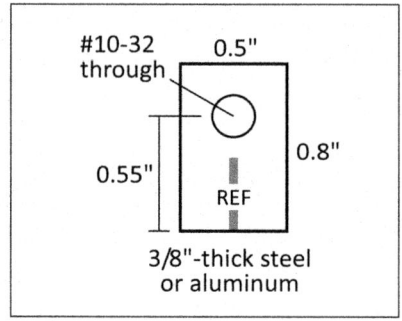

FIGURE 12-14   Transfer block.

line 0.275" from the left hand edge. Center-punch the spot; then drill #8-32 clear for the lead-screw post. The moment of truth comes when you dry-fit the working parts. The wedge should align perfectly with the T-square ramp; if not, all you can do is enlarge the locating hole for the lead-screw post.

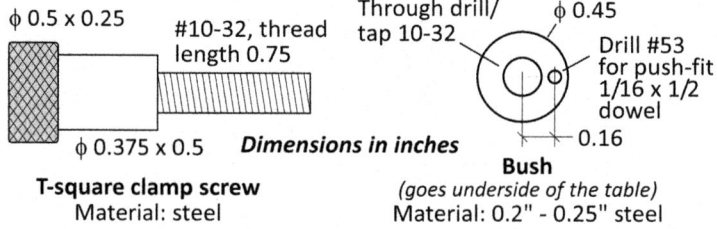

FIGURE 12-15   T-square clamp screw.

12-10 TOOLHOLDER

The toolholder (Figure 12-16) was made from 1-1/2"-square steel. The tool slot in my case has 1/2" clearance in both directions. The three clamp set screws are #10-32, on a line nearer the open end than the back of the slot. (*Why?* So they can be used in a pinch to hold round bar, e.g., a wheel dresser.) A brass shim washer on the underside allows the toolholder to slide freely when its angular position causes it to overhang the T-square wedge. The locking handle is similar to that used on the T-square.

FIGURE 12-16 Toolholder. The inset shows the circular brass shim washer, which is glued to the underside.

12-11 TOOL SLIDE

This was cut from 1/4" steel plate (Figures 12-17 and 12-18). The 2" slot is for the slide stop, used to prevent collisions between the grinding wheel and the tool being ground.

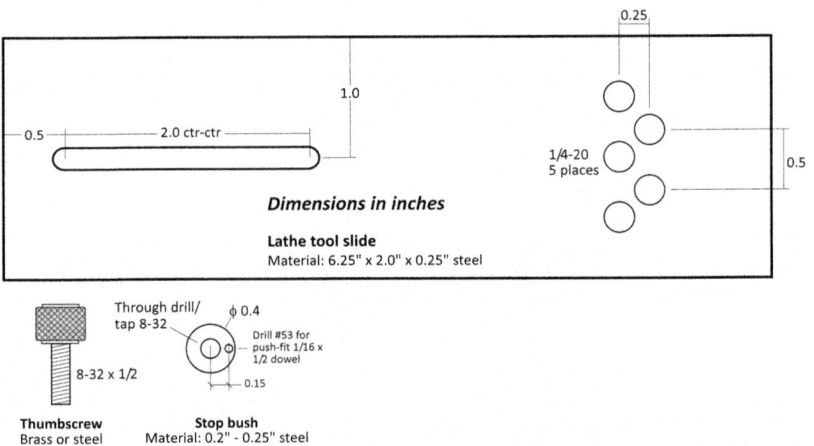

FIGURE 12-17 Tool slide and stop components.

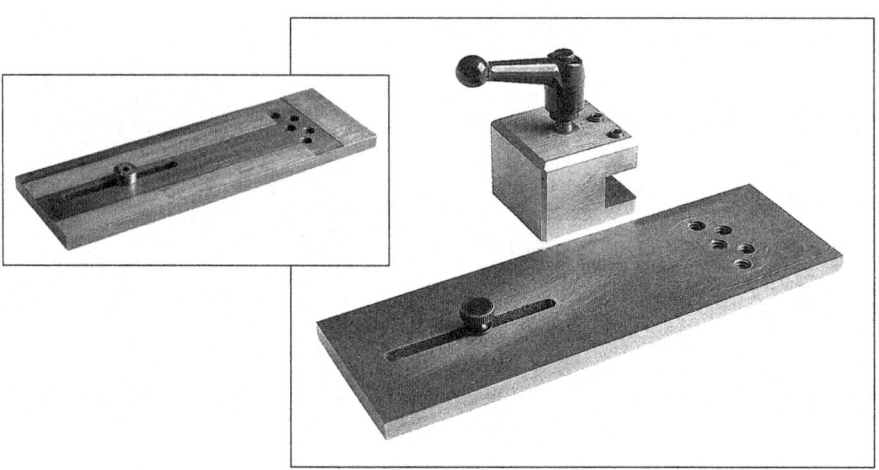

FIGURE 12-18 Tool slide. The inset shows the brass wear strips on the underside.

Note (in both figures) the pattern of five 1/4-20 holes on the lathe toolholder slide; this extends the range of fore-aft positioning—not often a factor, but great when you need it. The wear strips on the underside of the slide can be brass or plastic shim, say 0.01" thick, attached with spray adhesive.

Whatever else you do, don't apply grease or oil to any part of the grinder.

12-12 SAFETY GUARD

Since the photos were taken, the PVC baffle in front of the motor was replaced by a 6" x 6" x 1/8" aluminum sheet. This allowed me to add a 6" x 4" guard of clear 1/8" polycarbonate to cover the work area. This is not a substitute for safety goggles, *which should be worn for all grinding*, but it should lessen the danger of dust and/or airborne chunks of grinding wheel.

12-13 USING THE TOOL GRINDER

Job #1 is to dress the front-facing flat surface—the rim—of the grinding wheel (Figure 12-19). You should not need to touch the tapered surfaces.

FIGURE 12-19   Diamond dresser (1/2" shank). The dark ring on the wheel rim is glazing that will be removed by the diamond.

The wheel will need to be redressed from time to time, especially if glazing appears. Glazing looks just like it sounds—shiny, a surface that doesn't do anything but heat up the tool.

The ideal location for wheel dressing is outdoors, but most of us will settle for a shop vac instead, setting up the grinder as far as possible from other machines. You will need a single-point diamond dresser with a 1/2"-diameter shank. This can be installed in the lathe toolholder, provided the set screws are over center as described earlier. Tilt the table 5° to 10°, diamond dressing tool pointing upwards (Figure 12-19), then set the T-square so that the diamond just clears the wheel. Run the motor; then, pressing the tool slide firmly against the wedge, make left-right passes across the wheel, advancing the dresser in 0.001" increments until you are satisfied that the rim runs true. Be prepared for a lot of dust.

Table 12-1 shows the tool angles available with the grinder.

TABLE 12-1   Available tool angles

Knife Tools (left hand and right hand)	Grinder Capability (range in degrees)
Side rake angle	0 to 20
Back rake angle	Positive: 0 to 10
	Negative: any angle
Side relief angle	Any angle
End relief angle	0 to 20
End cutting edge angle	Any angle
Side cutting edge angle	0 to 30

12-14 GRINDING A LATHE TOOL FROM SCRATCH

Let's assume you will be starting with a square blank of HSS. This is a distant relative of the basic tool steel (e.g., W1 or O1) that you might use to make special cutters in the shop—simple end mills, D-bits, and counterbores, for instance. For one thing, HSS is not machinable in the ordinary sense, so it has to be ground to the desired shape. For another, it can be ground to very high temperatures (about to turn dull red) without affecting its performance in the slightest—a good thing if you are trying to grind, say, a single-point threading tool from a 3/8"-square blank of M2. That would take all day with the dainty technique we use for shop-hardened steels like O1 and W1 (which, once hardened, cannot tolerate even the slightest over-heating). Finally, no matter what you've read elsewhere, *don't cool HSS by water quenching.* Allow time for air cooling.

There is a choice of HSS alloys, M2 being one of the most basic. There are also several high-cobalt alloys, but these are harder to grind and cost at least 100% more than M2.

12-15 HOW LARGE A BLANK DO YOU NEED?

For less wear and tear on your grinding wheel, the smaller the better—it makes no difference, provided the cutting edge is keen and the angles are to spec. In theory, you could use a 1/8" square, but most people find 3/16" or 1/4" blanks more practical. In many cases (Figure 12-20), you can use a round blank salvaged from an HSS drill.

Think of your shop-built grinder as a precision finishing machine. Therefore, do the roughing out on a standard bench grinder with a lower-budget wheel. Otherwise, you will be doing more wheel dressing than you'd like.

FIGURE 12-20 Round 3/16" bit ground as right-hand knife tool. The shank is 1/2"-square steel. Before grinding its business end, make sure the round bit doesn't rotate by grinding a rough flat for the set screws.

12-16 GRINDING A KNIFE TOOL

Before you start grinding, install the tool to be worked on in the toolholder; then use a protractor to set the table at the desired angle (Figure 12-21). Set the table by reference to the tilt scale (shown earlier in Figure 12-4). Always do a dry run first, motor stationary, to be sure you can complete the grinding pass without interference; then set the slide stop to limit left-to-right travel. Press the tool slide firmly against the wedge throughout the grinding pass.

FIGURE 12-21 Set the toolholder angle.

Rough out the tool shape on a bench grinder; then go to the precision grinder for finishing. Figure 12-22 shows how to grind a general purpose knife tool for soft materials, with side rake, side relief, and end relief angles all 10°.

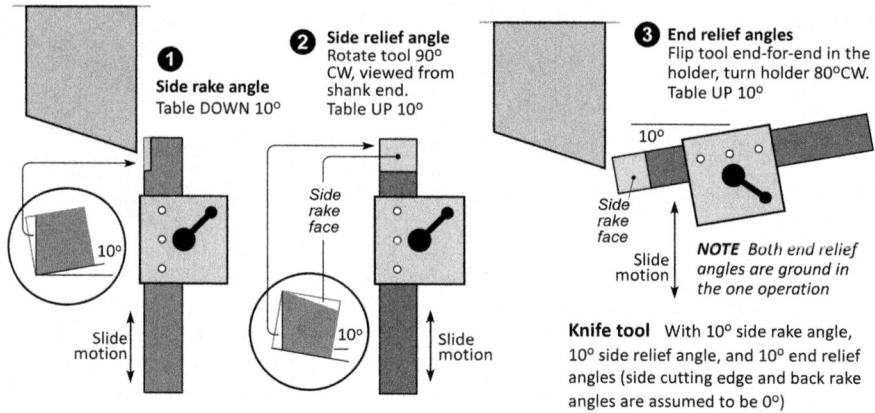

FIGURE 12-22   Grinding a RH knife tool.

Remove the tiny burrs left by the grinder using a fine stone or diamond lap, taking great care not to round over the edges. The slightest error here makes a world of difference, so hold the stone/lap firmly against the ground surface, and stroke gently without rocking. Finally, stone a tiny radius at the very tip of the tool, where the side angles and end angles meet. This can significantly improve surface finish.

12-17 GRINDING A THREAD-CUTTING TOOL

If you start with a square blank, a lot of material has to be ground off to achieve a 60° point. This should be done on a conventional bench grinder before precision grinding. While at the bench grinder, consider also thinning the tip of the blank to a tongue about 3/16" thick. This isn't strictly necessary, but it does reduce the amount of material to be removed when working on the 60° Vee. Grind the Vee as in Figure 12-23.

This shows side relief angles of 10°, good for most work, but you may prefer to decrease this a few degrees for better support at the cutting edge. The final step in making a threading tool, as with all other tools, is to hone the edges and flatten the tip crosswise (otherwise, it will snap off in use, taking with it more than just the tip). How much flattening is permissible depends on the thread pitch—for instance, less than 10 thousandths for 32 TPI.

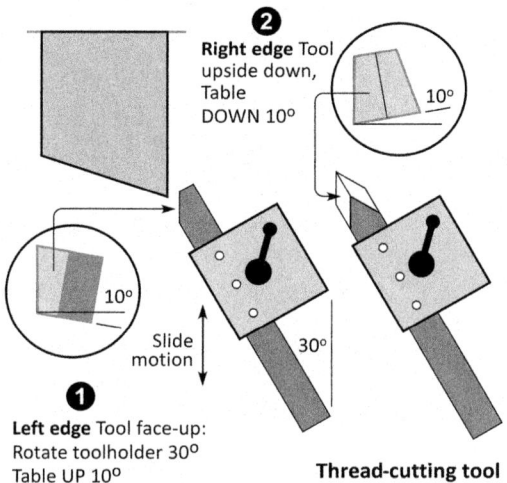

2
Right edge Tool upside down, Table DOWN 10°

10°

10°

Slide motion

30°

1
Left edge Tool face-up:
Rotate toolholder 30°
Table UP 10°

Thread-cutting tool

FIGURE 12-23   Grinding a thread-cutting tool.

12-18 DIAMOND GRINDING WHEEL OPTION

The aluminum oxide wheel described above is good for all tool steel grinding, but it cannot be used on carbide insert tools. For these, use a diamond wheel instead (Figure 12-24). You will probably need to machine a hub, probably aluminum, to suit the wheel and your particular motor.

The prices of diamond wheels from mainstream suppliers can be fairly shocking, but there are better buys on eBay, etc. (The one shown in Figure 12-24 cost about $30. It has worked well for more than 10 years.) *One note of caution:* Use the diamond wheel only when you have to. Don't wear it out on regular tool steels.

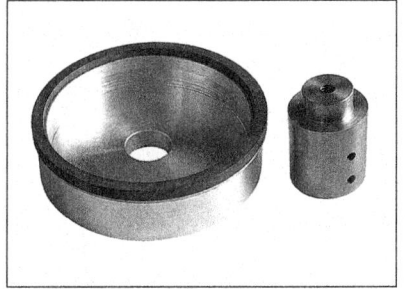

FIGURE 12-24   Typical 100-mm diamond-faced wheel.

Really New to All This?

APPENDIX AT A GLANCE

Once your lathe is in its working location, leveled and lubricated, the usual questions are *Where to start?* and *On what?*

You could start with something like the following experiment. The low-cost, low-risk workpiece shown in Figure A-1 demonstrates a number of everyday lathe routines: Three different *concentric* diameters, a partially threaded through hole, and a counterbore for a #10-32 socket head cap screw.

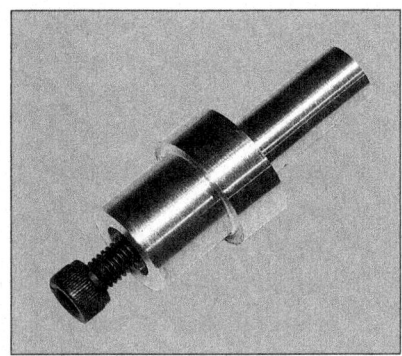

FIGURE A-1 A suggested first experiment.

Suggested dimensions are given in Figure A-2. Where additional information may be helpful, references are given to other sections of the book.

You will need the usual metal shop hand tools, such as:

- Hacksaw
- Vise, with jaw covers to prevent surface damage
- Files (Section 4-19)
- Vernier or digital caliper (nothing fancy; figure on $20 or so)
- Twist drills (hardware store sizes will do for this first experiment), 11/64" and 5/16"

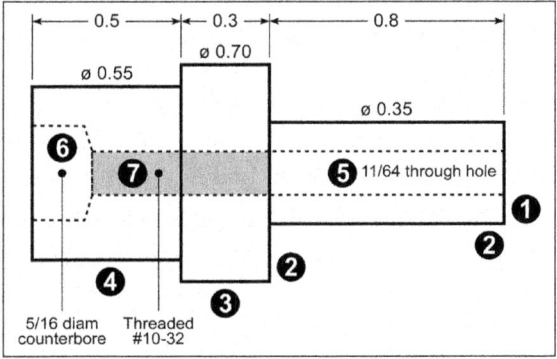

FIGURE A-2 Suggested dimensions in inches for each part of the workpiece shown in Figure A-1. The Greek letter phi, ϕ, is often used to denote diameter. The bottom of the counterbore is funnel-shaped, as drilled. An end mill would leave a flat bottom, which is preferred; see Section A-10.

- Optional, but you will definitely need one quite soon: #2 center drill (Section 2-28)
- #10-32 tap, standard plug chamfer (Section 2-38)

Aside from the above, you will need these lathe-specific items:

- 3-jaw chuck (often supplied with the lathe)
- Right-hand knife tool (Section 2-18)
- Tailstock drill chuck (Section 1-27)

For the workpiece you will need:

- 3/4"-diameter aluminum or (less expensive) Type 1 rigid PVC rod (McMaster Carr, Section 4-42). If your lathe has a hollow spindle with an ID larger than 3/4", there's no need to cut off a length for the workpiece. If your lathe spindle is solid, you *do* need to cut the rod. This can waste material, because you will need at least an additional inch, probably more, to hold the unworked portion in the chuck—a surplus length that will likely end in the shop recycle bin.

A-1 THE RH KNIFE TOOL

Instead of grinding your own tool from a high-speed steel blank (a lengthy operation), start with an off-the-shelf carbide-tipped tool. A tool you can use for practically every material is an indexable carbide insert (Section 2-18), the shank of which can be set at a slight angle to allow both face cutting and traverse cutting (right-to-left saddle motion), as noted in Section 3-2, Figure 3-4.

A-2 TOOLHOLDER

Your lathe probably came with a 4-tool turret, which you will probably want to replace quite soon with a Quick-Change Tool post (Section 1-28). Assuming you are starting with a 4-tool turret, the first job is to insert sheet metal scraps (shims) under the tool shank to set the cutting edge of the knife tool exactly at the height of the lathe's centerline. (If "exactly" is hard to gauge, slightly below the centerline is better than above.) At some later time, you will probably make a dedicated height gauge to rest on the cross-slide surface (Section 3-11). For today, install in the spindle the taper that probably came with the lathe, as shown in Figure A-3. (If not, use the tailstock center instead—swing the tool around to touch it.)

FIGURE A-3   Setting the tool height.

A-3   CLAMP THE MATERIAL

Install the 3-jaw chuck on the spindle. Be sure to fasten it securely (Section 1-14). Clamp the 3/4" material firmly in the 3 jaw, with about 1-1/2" exposed to the right of the jaws. Don't overdo the clamping pressure.

A-4 FIRST CUT: FACING THE WORKPIECE—STEP 1

Swing the knife tool 3° to 5° counterclockwise so it can be used for both face cutting and traverse cutting (diameter reducing). Tighten the tool-holder; then use the cross-slide and saddle controls to move the tool clear of the workpiece. With the lathe spindle running at about 500 rpm, move the tool *slowly to the left* until it just contacts the end of the workpiece (Figure A-4).

FIGURE A-4   Facing cut, (1) on Figure A-2.

Without moving the saddle, ease the tool tip toward the centerline by turning the cross-slide handle. This should cut a light shaving off the end face.

Retract the cross-slide, move the saddle left a few mils, and then make a second cut. Repeat the procedure until the face is fully machined. If a small nub of material remains at the centerline, raise the knife edge by shimming under the tool shank (if a conventional turret) or by adjusting the thumbnut (if a QCTP). Depending on the style of knife tool installed, it is usually possible to face-cut in either direction of the cross-slide, *IN* or *OUT* (out-feeding usually gives the better finish).

A-5 REDUCING THE DIAMETER—STEP 2

The outer end of the workpiece is to be 0.35" diameter and 0.8" long. If your lathe doesn't have digital scales (DRO), coat the material with marking fluid—the sort that comes in a little can with a brush. Let it dry; then set and lock your calipers at 0.8". Set one jaw on the outer face; then hand-turn the spindle to scribe the circumference of the rod with the other jaw (Figure A-5).

FIGURE A-5 Marking 0.8" from the face. *Note:* For all other shots of the workpiece on these pages, a DRO was used, so there was no need for the scribed marker shown here.

Using the cross-slide, back the knife tool well clear of the outer surface of the workpiece; then move it forward to *just* touch the outer surface of the workpiece at any point to the right of the scribed mark. (Outward motion, followed by forward, takes care of *backlash error* in the cross-slide—not necessary if you have a DRO.) Zero the cross-slide dial (or the X axis on the DRO). Move the saddle right to clear the tool; then infeed the cross-slide 20 or 30 divisions (0.02" to 0.03"). With the spindle again running at about 500 rpm, make the first traversing pass to the scribed line, moving very slowly when nearing the mark to avoid overshooting.

Make additional traversing passes with as much infeed each time as you are comfortable with. After the first couple of passes, check the diameter with the calipers; then look at the cross-slide dial reading. Calculate how many more divisions it will take to get to 0.35". (Remember, that's *diameter*; the infeed is *one-half* of that.)

Make the second-to-last cut a little short of the finished size. This should allow you to make a very light finishing cut (Figures A-6, A-7, and A-8).

FIGURE A-6 Diameter reducing—first pass.

FIGURE A-7   Diameter reducing—second pass.

FIGURE A-8   Finished ɸ 0.35" section, (2) on Figure A-2.

Aside from a DRO, there is usually no way to precisely gauge right-to-left motion of the saddle, aside from after-the-fact measurement with a caliper (use the "depth rod"). Cutting to a scribed line will give an accuracy of about ± 0.02", at best. However, even without a DRO, you can use the compound instead of the saddle to achieve much greater precision, say, ± 0.002". To do this, set the compound at *exactly* zero degrees; then refer to the compound's graduated dial when making the cutting passes. When you are close to the target diameter, check the result with the caliper depth rod; then adjust the compound travel, if necessary, when making the final cut to length.

A-6 TURNING THE LARGE DIAMETER—STEP 3

If you are working with 3/4"-diameter stock, this can be done in just two passes, one for roughing to size and the second for finishing. Any length of cut will do, provided it is comfortably over 0.3" (Figure A-9).

FIGURE A-9 Turning the 0.7" section, (3) on Figure A-2.

A-7 CUTTING OFF THE WORKPIECE

There are three ways to do this: (1) Take the entire rod to a bench vise; then cut off with a hacksaw. (2) Use a hacksaw with the rod clamped in the 3 jaw (*but protect the lathe bed with scrap wood;* see Section 1-11, Figure 1-11). (3) If more adventurous, use a cut-off blade with the spindle running more slowly than you use for standard turning (Sections 4-24 and 4-25).

Before doing any of the above, you might be wondering: *Why not machine the 0.55"-diameter section (4) before cutting off?* In this case, the answer is that the only tool we assumed to be available is a RH knife tool, which cannot do the job. However, if you do happen to have a LH knife tool, use it for the 0.55" section; then cut off the workpiece (see the "Concentricity" box).

CONCENTRICITY

This is an issue in many turning projects, because it can affect the entire process. For instance, in the experimental job described here, how important is it for the ϕ 0.35", ϕ 0.70", and ϕ 0.55" sections to be perfectly concentric? You might be hoping for concentricity within as little as 1/10 of a mil ($\pm$ 0.0001"). For Steps (2) and (3), no problem, but the ϕ 0.55" section (4) is another matter. It *could* be possible to achieve the same perfect concentricity for (4), but only if you *machine it in place* using a LH knife tool. On these pages, we describe the other way to do the job, flipping the workpiece end-for-end. This is what machinists do when concentricity isn't critical, but to be OK with "flipping," they need to be sure of reasonable 3-jaw accuracy (Chapter 7), or are prepared to take the time to adjust a 4 jaw for minimum runout.

In the following sections, we press on with just the RH knife tool.

A-8 FLIPPING THE WORKPIECE—STEP 4

An assumption here is that your 3-jaw chuck is "reasonably accurate," meaning that it gives a TIR (total indicator reading) not greater than, say, 0.005" when holding a truly round section—in this case, the ϕ 0.35" section (Figure A-10). For more on TIR, see Section 4-30.

FIGURE A-10   Workpiece flipped end for end.

Machine and measure the ϕ 0.55" section (4) on Figure A-2, in the same way as you did the ϕ 0.35" section.

For a reliable way to get better concentricity of the flipped workpiece, use a 4-jaw chuck in place of the 3 jaw (Section 4-33). It takes only a few minutes to set it up for a TIR of less than ± 0.001".

A-9 11/64 THROUGH HOLE

If you have a center drill, install it in the tailstock chuck; then run it in a short distance (say, 1/8") to make a starter hole in the workpiece (Figure A-11). If you don't use a center drill, the 11/64" drill in the next operation will possibly start off-center, then wobble.

FIGURE A-11   Center drilling.

Exchange the center drill for the 11/64" drill; then drill the through hole (Figure A-12). Cut-and-try experimenting will tell you the best spindle speed and tool pressure for this. Two other variables to bear in mind are *workpiece material* and *drill sharpness*.

FIGURE A-12   Drilling the 11/64"-diameter through hole.

This is a deep hole, so you will need to withdraw and reinsert the drill many times, even if you are using cutting fluid (Section 4-9).

For a timesaving way to do this, *apply pressure by sliding the tailstock to the left*, instead of cranking the handwheel.

The 11/64" (0.172") through hole will later be tapped #10-32. If we were going by the book, the drill for that tap size would be #21 (0.159"), a difference of 13 mil. The 11/64" drill was chosen because that's the nearest size you'll find in the typical home shop or home store.

You might be wondering: *Does the difference in hole size matter?* Not really. You lose very little strength in the thread with a larger-than-nominal hole. And there is much less danger of breaking the tap (Section 2-39).

A-10 DRILLING THE COUNTERBORE—STEP 6

Exchange the 11/64" drill for a 5/16" drill (Figure A-13). Lower the spindle speed by about 50%, and then run the drill in about 0.3". (Counterbore depth is usually not critical, usually just enough to sink the head of a screw or other component.) This leaves a funnel shape at the bottom of the counterbore, which is not ideal. However, it could be flattened in just a few seconds with a 5/16"-diameter end mill. This would need to be a *single-end* cutter that can be held in the tailstock chuck.

FIGURE A-13   The 5/16"-diameter counterbore, (6) on Figure A-2.

A-11 TAPPING THE THROUGH HOLE—STEP 7

Depending on the style of tap wrench available, there are several ways of doing this. No matter how you do it, the overriding requirement is to keep the tap perfectly on the lathe's centerline throughout. For tap sizes #10-32 or M5 and larger, you may not need a tap wrench to start the thread. Instead, grip the tap in the chuck, and allow the tailstock to slide on the bed; then, while applying light leftward pressure by pushing on the tailstock, crank the chuck round by hand, five or so complete turns (Figure A-14).

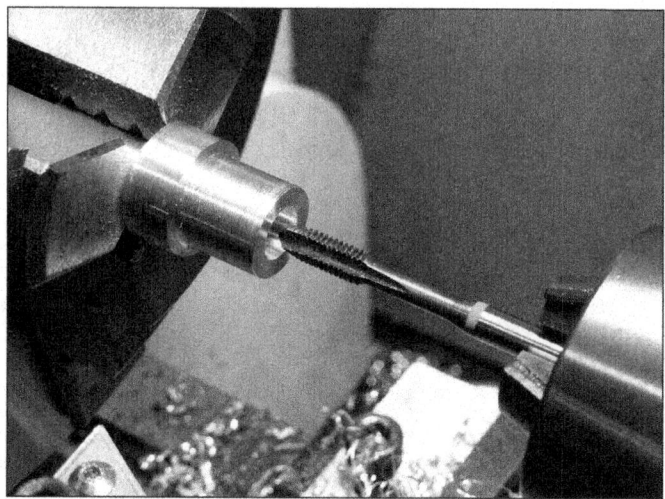

FIGURE A-14   The #10-32 tap.

Then, having ensured solid alignment, finish the job using a tap wrench (Figure A-15). *Caution!* Could your nicely-finished ϕ 0.35" section be marred by slipping in the chuck? If so, take the workpiece to a bench vise, with protective (e.g., leather) pads.

FIGURE A-15   Finishing with a tap wrench (7) on Figure A-2. This "open-style" wrench cannot be used for initial threading, because there is no reliable way to keep the tap on the centerline.

Caution: Using the tailstock chuck as a holding device is *not good* for smaller taps, because drag from the tailstock is likely to break the thread and/or the tap. Instead, use a sliding tap holder such as described in Section 2-40, or use the more conventional type shown in Figure A-16. Chuck wrenches like this are usually center-drilled as an aiming point for the tailstock center.

FIGURE A-16 Another type of tap wrench. This chuck-style wrench is good for sizes up to 1/4". The shank is center-drilled for alignment.

To make sure the tap is held precisely on the lathe centerline, keep the tailstock center engaged in the hole by advancing the tailstock barrel (turn the handwheel) after every few degrees of rotation of the workpiece.

Threading under power on the lathe is certainly possible—routine for some users—but is *not recommended*, even for taps larger than #10.

Index

References to figures are in italics.

About the Author

Richard Rex has worked on lathes and milling machines since his teen-years in a home shop, and later on a variety of production machines (his current home shop setup has a 12" x 36" lathe and a Bridgeport mill). More recently, he has set up several engineering lab model shops from scratch, with the usual complement of Hardinge lathes and Bridgeport mills.

Richard worked for 10 years in product marketing management with Hewlett Packard and Brown Boveri in the UK. In the US, he has been CEO of several engineering/manufacturing companies. He has been a "tech writer" throughout his working career, starting with monographs on signal processing with HP, later (as CEO) data sheets and application notes for a wide range of his company's products. In recent years he has written and illustrated 30+ manuals and tech bulletins for a machine tool distributor in Pittsburgh.